INTERNATIONAL CENTRE FOR MECHANICAL SCIENCES

COURSES AND LECTURES - No. 63

PETER Chr. MÜLLER

TECHNICAL UNIVERSITY OF MUNICH

SPECIAL PROBLEMS OF GYRODYNAMICS

COURSE HELD AT THE DEPARTMENT
OF GENERAL MECHANICS
OCTOBER 1970

UDINE 1970

SPRINGER-VERLAG WIEN GMBH

Originally published by Springer-Verlag Wien New York in 1972

ISBN 978-3-211-81085-9   ISBN 978-3-7091-2882-4 (eBook)
DOI 10.1007/978-3-7091-2882-4

Preface

This textbook contains my lectures given at the International Centre for Mechanical Sciences (CISM) in Udine, Italy, in October 1970. These lecture notes are intended as a supplement to the course "Gyrodynamics" of Professor Dr.K.Magnus held at CISM at the same session. Therefore, some familiarity with the fundamentals of gyrodynamics as an introductory level is presupposed on the part of the reader.

The book is divided into three main parts. The first part (Chapter 1) is concerned with the shape triangle which is a basic tool for the presentation of stability regions of rotating rigid bodies, where stability depends substantially on the principal moments of inertia. For illustration, stability regions of permanent rotations of a heavy unsymmetrical body are discussed by application of the shape triangle. The purpose of Chapter 2 is to give a general idea of stability methods, specifically provided for mechanical systems. As a result, the famous THOMSON - TAIT stability theorem is extended to pervasive damped systems by means of the modern concept of controllability. For example the stability of a two-body-satellite on circular orbit is investigated. The third part comprises chapters 3 and 4 which are devoted to error analysis and error propagation as well as self-alignment tech-

niques of inertial platforms. An inertial platform is the centre of an inertial guidance system and its errors substantially influence the quality of any navigation. The errors come from gyro drifts, accelerometer bias, unorthogonality of the gyro and accelerometer input axes, misalignment of the platform, etc. In the longtime error propagation, misalignments, besides gyro drifts, are the most troublesome errors. Therefore, linear and nonlinear alignment methods are discussed in detail and a time-optimal alignment loop is proposed.

In conclusion, I wish to express my gratitude to both the International Centre for Mechanical Sciences and Professor Dr.K.Magnus who encouraged me to do this work. I am also indebted to my colleagues at the Institute of Mechanics, Technical University of Munich for their helpful suggestions and constructive criticisms.

P. C. Müller.

*niques of inertial platforms. An inertial platform is
the centre of an inertial guidance system and its errors substantially influence the quality of any navigation. The errors come from gyro drifts, accelerometer bias, unorthogonality of the gyro and accelerometer input axes, misalignment of the platform, etc.
In the longtime error propagation, misalignments, besides gyro drifts, are the most troublesome errors.
Therefore, linear and nonlinear alignment methods are
discussed in detail and a time-optimal alignment loop
is proposed.*

      *In conclusion, I wish to express my
gratitude to both the International Centre for Mechanical Sciences and Professor Dr.K.Magnus who encouraged me to do this work. I am also indebted to my
colleagues at the Institute of Mechanics, Technical
University of Munich for their helpful suggestions
and constructive criticisms.*

                                    P. C. Müller.

Preface

This textbook contains my lectures given at the International Centre for Mechanical Sciences (CISM) in Udine, Italy, in October 1970. These lecture notes are intended as a supplement to the course "Gyrodynamics" of Professor Dr.K.Magnus held at CISM at the same session. Therefore, some familiarity with the fundamentals of gyrodynamics as an introductory level is presupposed on the part of the reader.

The book is divided into three main parts. The first part (Chapter 1) is concerned with the shape triangle which is a basic tool for the presentation of stability regions of rotating rigid bodies, where stability depends substantially on the principal moments of inertia. For illustration, stability regions of permanent rotations of a heavy unsymmetrical body are discussed by application of the shape triangle. The purpose of Chapter 2 is to give a general idea of stability methods, specifically provided for mechanical systems. As a result, the famous THOMSON – TAIT stability theorem is extended to pervasive damped systems by means of the modern concept of controllability. For example the stability of a two-body-satellite on circular orbit is investigated. The third part comprises chapters 3 and 4 which are devoted to error analysis and error propagation as well as self-alignment tech-

Chapter 1
The Shape - Triangle

1. 1. Introduction

The stability of gyroscopic devices or satellites often only depends on the mass geometry of these bodies, i.e. on their moments of inertia. Thus, a gravity-gradient stabilized satellite on a circular orbit around the earth has a stable position only if the principal moments of inertia A, B, C satisfy the condition

$$B > A > C \qquad (1.1.1)$$

or the conditions

$$A > C > B \qquad (1.1.2a)$$

and

$$\left[1 + 3 \frac{B-C}{A} - \frac{B-C}{A} \frac{A-B}{C} \right]^2 + 16 \frac{B-C}{A} \frac{A-B}{C} > 0 \qquad (1.1.2b)$$

(notation and detailed discussion see [1]).Relation (1.1.2b) especially shows that the statement of stability conditions requires a suitable presentation of the moments of inertia. Fundamentally the mass geometry can be described in a three dimensional parameter space where the principal moments of inertia are the ordi-

nates of a set of rectangular axes. But there are two disadvan-
tages ; firstly, the nonplanar description is cumbersome and un-
illustrative and, secondly, this description bears no relation
to the special type of the stability conditions. The degree of
stability does not depend on A , B , C but only on the ratio
$A : B : C$. There are two essential parameters only and therefore the stability regions can be registered in two -dimensional
parameter spaces.

There are various presentations of mass geometry of a rigid body
$[2, 3]$. According to SCHULER the moments of inertia are inter-
preted as the three straight lines of a triangle (Fig.1a), whilst
GRAMMEL regards the ratios $B : A$ and $C : A$ as coordinates of two
rectangular axes (Fig.1b). KANE introduces the ratios $(A - B)/C$
and $(B - C)/A$ as stability patameters (Fig.1c) and finally,
MAGNUS presents the moments of inertia in the so-called shape
triangle (Fig.1d). The four different theories are topologically
equivalent but only the description of the three principal moments of inertia in the shape triangle is symmetrical with respect to all moments. A detailed discussion of the applicability
of the shape triangle is therefore presented in this lecture
note.

1. 2. Fundamentals of the shape triangle

The shape of a rigid body — more exactly the shape
of its momental ellipsoid — is characterized by the principal

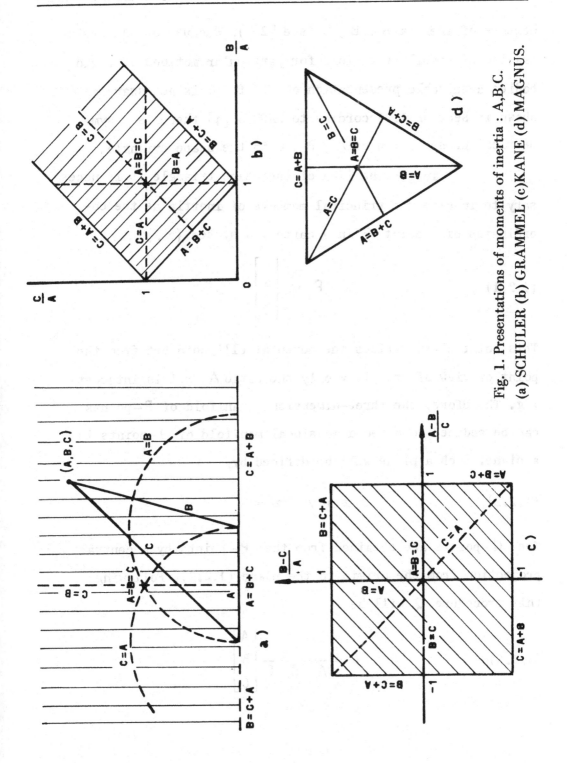

Fig. 1. Presentations of moments of inertia : A,B,C.
(a) SCHULER (b) GRAMMEL (c)KANE (d) MAGNUS.

moments of inertia A , B , C (see [2]). Because of the regis-
tration of stability regions for particular motions of rigid
bodies a suitable presentation of A , B , C is necessary as
shown in section 1. According to MAGNUS [4] there is a symmetrical presentation of A , B , C in the shape triangle.

To get the idea of the shape triangle it is neces
sary to imagine the principal moments of inertia as the coordinates of a point P in a cartesian system $\{x, y, z\}$:

$$(1.2.1) \qquad\qquad P = \begin{bmatrix} A \\ B \\ C \end{bmatrix} .$$

This point characterizes the momental ellipsoid but from the
point of view of stability only the ratio $A:B:C$ is interesting. Therefore, the three–dimensional manifold of P–points
can be reduced to a two–dimensional manifold of Q–points in
a plane. Such a plane will be defined by

$$(1.2.2) \qquad\qquad x + y + z = 1 .$$

The Q–points are obtained from the P–points by a central
projection from the origin to the plane (1.2.2). Following
this procedure Q will be

$$(1.2.3) \qquad\qquad Q = \frac{1}{A + B + C} \begin{bmatrix} A \\ B \\ C \end{bmatrix} .$$

Because of the lack of negativity of A, B, C the Q's are con-
tained in a fundamental triangle „A", „B", „C" (Fig. 2). With respect to the triangle's inequalities

$$A \leq B + C, \quad B \leq C + A, \quad C \leq A + B \qquad (1.2.4)$$

the possible Q's are restricted to the triangle a, b, c
(Fig. 2). This is the shape triangle.

The presentation of the shape of momental ellip-
soids in the shape triangle can be characterized as follows :

. Corners correspond to bars,

. side lines correspond to discs,

. middle lines correspond to symmetrical bodies,

. the intersection of the middle lines (middle
  point) correspond to a momental sphere,

. every part-triangle, bordered by two middle
  lines and one side line is characterized by a
  relation of the moment of inertia as shown in
  Fig. 3.

This method of dividing the shape triangle into
sections however, is not sufficient for the calculation of stability boundaries. For example, what region in the shape triangle
may be given by the stability conditions (1.1.2) ? The problem is
therefore concerned chiefly with the calculation of Q-points
of the shape triangle.

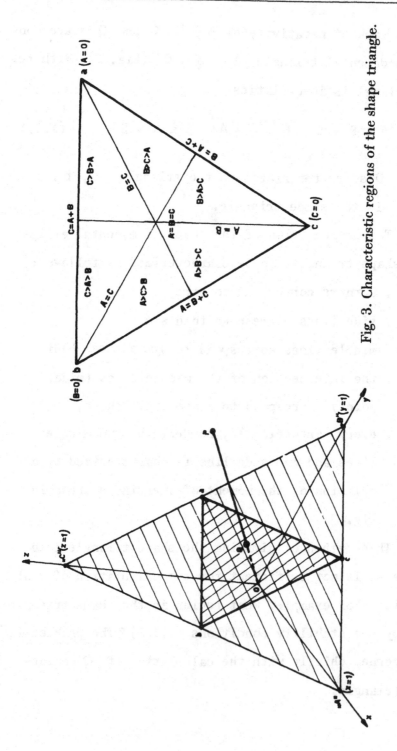

Fig. 3. Characteristic regions of the shape triangle.

Fig. 2. Fundamental triangle "A", "B", "C" and shape triangle abc.

1. 3. Determination of Q-points in the shape triangle

Mathematically speaking the Q-points are described by barycentric coordinates (A , B , C) which means that Q is the centre of mass if in the corners „A" , „B" , „C" of the fundamental triangle three "masses" A , B , C , respectively are fixed. This characterization of Q however is not a simple method of calculating Q . Four methods for the determination of Q are described below.

1. 3. 1. Projected coordinates

From (1.2.3) it follows that two coordinates of Q must be chosen as stability parameters. For example,

$$h = \frac{A}{A + B + C} \quad , \quad 0 \le h \le \frac{1}{2} \quad , \qquad (1.3.1a)$$

$$k = \frac{B}{A + B + C} \quad , \quad 0 \le k \le \frac{1}{2} \quad , \qquad (1.3.1b)$$

$$1 - h - k = \frac{C}{A + B + C} \quad , \quad \frac{1}{2} \le h + k \le 1 \quad , \qquad (1.3.1c)$$

hence

$$Q \quad = \quad \begin{bmatrix} 0 \\ 0 \\ 1 \end{bmatrix} + h \begin{bmatrix} 1 \\ 0 \\ -1 \end{bmatrix} + k \begin{bmatrix} 0 \\ 1 \\ -1 \end{bmatrix} \quad . \qquad (1.3.2)$$

A fixed h and a variable k lead to a straight line parallel
to the side line opposite to the corner a (a corresponds to A
and, therefore, to h). The coordinate lines are parallels of
the side lines. Therefore, the construction of Q has to be done
as demonstrated in Fig. 4. Two ratios of $A/(A+B+C)$, $B/(A+B+C)$,
$C/(A+B+C)$ are calculated and marked on the linear scales
$(0.....0,5)$ on the side lines. Drawing the corresponding paral
lels, the intersection point will be Q .

1. 3. 2. Coordinates according to MAGNUS [4]

Often the ratios of the moments of inertia are
chosen as stability parameters. Therefore, it is convenient to
look for a coordinate system of these parameters. For example,
we select

(1.3.3a) $d = \dfrac{A}{B}$, $0 \leq d \leq \infty$,

(1.3.3b) $e = \dfrac{C}{A}$, $0 \leq e \leq \infty$,

(1.3.3c) $\dfrac{1}{ed} = \dfrac{B}{C}$, $0 \leq ed \leq \infty$.

The presentation of Q is now

(1.3.4) $Q = \begin{bmatrix} 0 \\ 0 \\ 1 \end{bmatrix} + \dfrac{1}{1+d+de} \begin{bmatrix} d \\ 1 \\ -(1+d) \end{bmatrix}$.

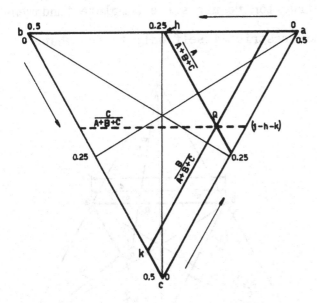

Fig.4. Construction of Q
by projected coordinates.

The manifold of Q's for a fixed d and a variable e is given
by a straight line which passes the corner „C″ of the fundamental triangle. Generally, such a coordinate line passes the
corner relating to that moment which does not appear in the
fixed ratio. The intersection point of the coordinate line with
the related side line of the shape triangle gives a scale-point,
e.g. for Q given by (1.3.4) the intersection with e = 0 leads
to

$$Q_d = \frac{1}{d+1} \begin{bmatrix} d \\ 1 \\ 0 \end{bmatrix}$$

(1.3.5)

which means that the scale is nonlinear with $sd/(d+1)$ where
s is the length of one triangle side. In Fig. 5 the construction of Q is shown as the intersection point of such two coor-

dinate lines. This construction requires the complete fundamental triangle and does not restrict itself only to the shape
triangle.

Fig.5. Construction of Q
by MAGNUS' coordinates.

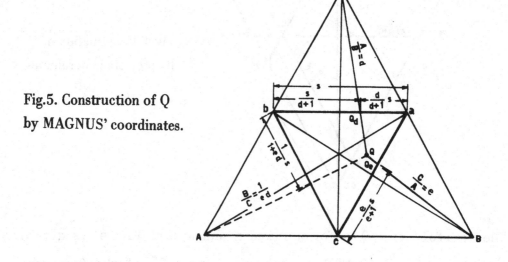

1. 3. 3. Coordinates according to SCHIEHLEN [5, 1]

In [5] SCHIEHLEN proposed coordinates which are
often required in the theory of satellites (see also [1]).
They are parameters such as the following :

(1.3.6a) $\qquad f = \dfrac{A - B}{C}$, $-1 \leq f \leq 1$,

(1.3.6b) $\qquad g = \dfrac{C - A}{B}$, $-1 \leq g \leq 1$,

(1.3.6c) $\qquad -\dfrac{f + g}{1 + fg} = \dfrac{B - C}{A}$, $-1 \leq -\dfrac{f + g}{1 + fg} \leq 1$.

The calculation of Q gives

$$Q = \frac{1}{2}\begin{bmatrix}1\\1\\0\end{bmatrix} + \frac{1+g}{3+fg+g-f}\begin{bmatrix}-\frac{1}{2}(1-f)\\-\frac{1}{2}(1+f)\\1\end{bmatrix}. \qquad (1.3.7)$$

The manifold of Q's for constant f and variable g is linear.
It is a line through the corner c which intersects the opposite
side line $(g=1)$ in

$$Q_f = \frac{1}{4}\begin{bmatrix}1+f\\1-f\\2\end{bmatrix}. \qquad (1.3.8)$$

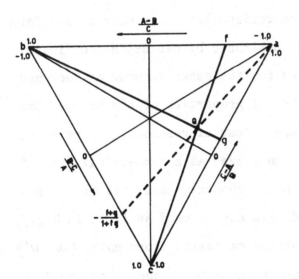

Fig.6. Construction of Q
by SCHIEHLEN's coordi-
nates.

Therefore, f itself is the linear scale factor on this line.
The construction of Q is the following (see Fig. 6) : Take two
ratios of $(B-C)/A$, $(C-A)/B$, and $(A-B)/C$ and mark them

on the linear scales of those side lines which correspond to the
divisors of the ratios. The connecting line of such a scale-
point and of the opposite corner represents a coordinate line.
Two independent coordinate lines will intersect in Q . The for-
mula (1.3.7) of Q may be difficult but the construction of Q
is very simple.

1. 3. 4. Cartesian coordinates

The stability boundaries are often computed by digital or analogue computer. Then it is convenient to draw the
boundary curves by means of an automatic plotter.Such instruments
are prepared for diagrams in rectangular coordinates. Therefore,
a characterization of Q is desirable by cartesian coordinates.
There are many possibilities for the establishment of rectangular coordinates. Any symmetrical properties of the problem can
be utilized in order to obtain clear expressions. For example,
if the middle line $B = C$ is envisaged as a symmetrical line of
a stability region, this line is chosen as one axis of the coordinate system. The second axis may be used as it is in Fig.7.
The point a is the origin of the cartesian coordinates $\{u', v'\}$
and $\{u, v\}$ where u and v are related to u' and v' by the distance s of two corners of the shape triangle. The relations to
the moments of inertia are given by

$$(1.3.9a) \qquad u = \frac{u'}{s} = -\sqrt{3}\ \frac{A}{A + B + C}\ ,$$

$$v = \frac{v'}{s} = \frac{C - B}{A + B + C} \quad . \tag{1.3.9b}$$

Q   becomes

$$Q = \begin{bmatrix} -\dfrac{1}{\sqrt{3}} u \\[2mm] \dfrac{1}{2}\left(1 + \dfrac{u}{\sqrt{3}} - v\right) \\[2mm] \dfrac{1}{2}\left(1 + \dfrac{u}{\sqrt{3}} + v\right) \end{bmatrix} \quad . \tag{1.3.10}$$

An easy geometrical construction of Q is not available. Therefore, the relations to the geometrical approaches of sections
1.3.1. – 1.3.3 are listed below :

a) Relations to the projected coordinates :

$$\left. \begin{aligned} u &= -\sqrt{3}\, h \quad , \\[2mm] v &= 1 - h - 2k \quad , \end{aligned} \right\} \tag{1.3.11}$$

$$\left. \begin{aligned} h &= -\frac{u}{\sqrt{3}} \quad , \\[2mm] k &= \frac{1}{2}\left(1 + \frac{u}{\sqrt{3}} - v\right) \quad . \end{aligned} \right\} \tag{1.3.12}$$

b) Relations to MAGNUS' coordinates :

$$\left. \begin{aligned} u &= -\sqrt{3}\,\frac{d}{1 + d + de} \quad , \\[2mm] v &= \frac{de - 1}{1 + d + de} \quad , \end{aligned} \right\} \tag{1.3.13}$$

$$(1.3.14) \begin{cases} d = \dfrac{2u}{\sqrt{3}\,v-(u+\sqrt{3})} \quad , \\[4mm] \dfrac{1}{e} = \dfrac{-2u}{\sqrt{3}\,v+(u+\sqrt{3})} \quad . \end{cases}$$

c) Relations to SCHIEHLEN's coordinates :

$$(1.3.15) \begin{cases} u = \dfrac{-\sqrt{3}\,(1+fg)}{3+fg+g-f} \quad , \\[4mm] v = \dfrac{f+g}{3+fg+g-f} \quad , \end{cases}$$

$$(1.3.16) \begin{cases} f = \sqrt{3}\;\dfrac{v-(\sqrt{3}\,u+1)}{\sqrt{3}\,v+(u+\sqrt{3})} \quad , \\[4mm] g = -\sqrt{3}\;\dfrac{v+(\sqrt{3}\,u+1)}{\sqrt{3}\,v-(u+\sqrt{3})} \quad . \end{cases}$$

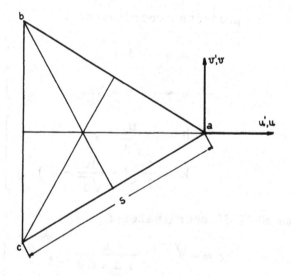

Fig.7. Cartesian coordinates.

1. 4. Examples

1. 4. 1. Gravity-gradient stabilized satellites

In [1] the problem of gravity-gradient stabilized satellites on circular orbits is discussed. With the notation of [1] the conditions of stability are given by (1.1.1) and (1.1.2). The moments of inertia A , B , C are the moments around the roll-, pitch, and yaw-axes, respectively. The stability regions are given by [1, Fig. 2.6] as shown in Fig. 8. In this example only some marked points Q_i ($i = 1,\ldots, 4$) of the boundary curve of (1.1.2b) are calculated. The Q_i's are characterized by the related equation of (1.1.2b) and the following properties :

$$Q_1 : \qquad C = A + B , \qquad\qquad (1.4.1a)$$

$$Q_2 : \qquad A = B , \qquad\qquad (1.4.1b)$$

$$Q_3 : \qquad C = A , \qquad\qquad (1.4.1c)$$

$$Q_4 : \qquad A = B + C . \qquad\qquad (1.4.1d)$$

The ratios of moments of inertia are determined by (1.1.2b) and (1.4.1). That gives

$$Q_1 : \frac{B-C}{A} = -1, \quad \frac{C-A}{B} = 1, \quad \frac{A-B}{C} = 10 - \sqrt{96} \approx 0,202 ; \quad (1.4.2a)$$

$$(1.4.2b) \quad Q_2 : \quad \frac{A-B}{C} = 0, \quad \frac{B-C}{A} = -\frac{1}{3}, \quad \frac{C-A}{B} = \frac{1}{3};$$

$$(\ Q_2 \text{ is double root of } (1.1.2b) \ ;$$

$$(1.4.2c) \quad Q_3 : \quad \frac{C-A}{B} = 0, \quad \frac{A-B}{C} \approx 0,161, \quad \frac{B-C}{A} \approx -0,161 \ ;$$

$$(1.4.2d) \quad Q_4 : \quad \frac{C-A}{B} = -1, \quad \frac{A-B}{C} = 1, \quad \frac{B-C}{A} = \sqrt{6} - \frac{5}{2} \approx -0,050 \ .$$

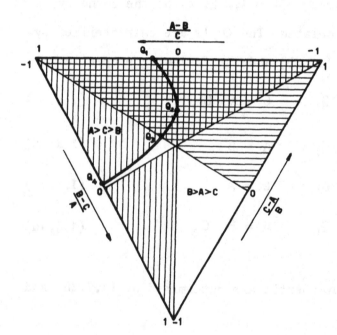

Fig.8. Stability diagram of gravity-gradient stabilized satellites on circular orbits. The shaded region is unstable.

The stability boundary (1.1.2b) is a curve of 4th order. A simple discussion is not possible ; therefore, a detailed calculation is better left to a digital computer.

1. 4. 2. Stability of the permanent rotations of a heavy, unsymmetrical rigid body about the vertical axis

The motion of a heavy, unsymmetrical rigid body about a fixed

point O is described by the Euler-equation

$$\dot{H} + \omega \times H = s \times G \qquad (1.4.3)$$

and the kinematical equation

$$\dot{a} + \omega \times a = 0 \ , \qquad (1.4.4)$$

where H is the angular momentum of the body with respect to the

fixed suspension point O, ω the angular velocity of the body,

s the vector from O to the centre of mass S, and

$$G = -mg a \qquad (1.4.5)$$

the gravity of the body in direction of the vertical unit vector

a . The derivatives refer to the body fixed axes.

An analytical solution of (1.4.3) and (1.4.4) is not

possible in general. But there are however a few particular cases

of integrability [2, 3], that of STAUDE [6] for instance being a

case in point. He showed that the above equations admit an ∞^1 of

solutions if no assumptions are made concerning the moments of

inertia and the position of the mass centre S of the body. These

solutions correspond to rotations of the body about the axes,

fixed in the body and in space. The vertical axis is one of the

permanent axes of rotation. These axes of rotation form a cone of the second order, concentric with the ellipsoid of inertia of the body at the fixed point 0 , but not coaxial with the latter and depending on the position of the mass centre S with respect to the principal axes of inertia of the body at 0 .

The condition for STAUDE's permanent rotations is

$$(1.4.6) \qquad (\omega \times a)_{t=0} = 0 \quad .$$

Then the solution can be constructed by means of the three classical integrals

$$(1.4.7) \quad \frac{1}{2} \omega \cdot H + mgs \cdot a = E_0 \qquad \text{(energy integral)},$$

$$(1.4.8) \quad H \cdot a = H_v \qquad \text{(integral of angular momentum)},$$

$$(1.4.9) \quad a \cdot a = 1 \qquad \text{(kinematical integral)},$$

and by STAUDE's integral

$$(1.4.10) \qquad a = a_0 \qquad \text{(body fixed vector !)}.$$

Now we will discuss the stability of the following permanent rotation

$$(1.4.11) \qquad \omega = \begin{bmatrix} \omega_0 \\ 0 \\ 0 \end{bmatrix}$$

of a heavy rigid body with the position

$$s = \begin{bmatrix} s \\ 0 \\ 0 \end{bmatrix} \qquad (1.4.12)$$

of the mass centre S. The solution of (1.4.3) to (1.4.6) is

given by

$$\omega(t) = \begin{bmatrix} \omega_0 \\ 0 \\ 0 \end{bmatrix}, \quad a(t) = a_0 = \begin{bmatrix} 1 \\ 0 \\ 0 \end{bmatrix}. \qquad (1.4.13)$$

The stability of the permanent rotation (1.4.13) will be inves-

tigated by the method of first approximation. A perturbed motion

of (1.4.13) may be stated by

$$\tilde{\omega} = \begin{bmatrix} \omega_0 + x_1 \\ x_2 \\ x_3 \end{bmatrix}, \quad \tilde{a} = \begin{bmatrix} 1 - x_4 \\ x_5 \\ x_6 \end{bmatrix}. \qquad (1.4.14)$$

By [3] the linearized model of the motion of (1.4.14) is given

by

$$x_1 = x_{10}, \quad x_4 = x_{40}, \qquad (1.4.15)$$

$$B\dot{x}_2 - (C - A)\omega_0 x_3 - mgs\,x_6 = 0, $$

$$\qquad (1.4.16a)$$

$$C\dot{x}_3 - (A - B)\omega_0 x_2 + mgs\,x_5 = 0, $$

$$(1.4.16b) \quad \begin{cases} \dot{x}_5 + x_3 \quad\quad\quad - \omega_0 x_6 = 0 \ , \\ \\ \dot{x}_6 - x_2 + \omega_0 x_5 \quad\quad = 0 \ . \end{cases}$$

The characteristic equation of (1.4.16) goes

$$(1.4.17) \quad\quad \left(\frac{\lambda}{\omega_0}\right)^4 + S_1 \left(\frac{\lambda}{\omega_0}\right)^2 + S_2 = 0$$

where

$$(1.4.18) \quad S_1 = \left(2 - \frac{A}{B} - \frac{A}{C} + \frac{A}{B}\frac{A}{C}\right) - k\left(\frac{A}{B} + \frac{A}{C}\right) \ ,$$

$$(1.4.19) \quad S_2 = \left(1 - \frac{A}{B} + k\frac{A}{B}\right)\left(1 - \frac{A}{C} + k\frac{A}{C}\right)$$

with

$$(1.4.20) \quad\quad\quad k = \frac{mgs}{A\omega_0^2}$$

as the ratio of potential and kinetic energy.
The permanent rotations (1.4.13) are stable if the roots of
(1.4.17) are imaginary. With the abbreviations

$$x = \frac{A}{B} \quad , \quad\quad y = \frac{A}{C}$$

the stability conditions are

$$S_1(x, y; k) \equiv (2 - x - y + xy) - k(x + y) > 0, \quad (1.4.21)$$

$$S_2(x, y; k) \equiv (1 - x + kx)(1 - y + ky) > 0, \quad (1.4.22)$$

$$S_3(x, y; k) \equiv S_1^2 - 4 S_2$$

$$\equiv [(2 - x - y + xy) - k(x + y)]^2 -$$

$$- 4(1 - x + kx)(1 - y + ky) \geq 0. \quad (1.4.23)$$

The ratios x and y are of the type of MAGNUS-coordinates. With regard to the triangle inequalities (1.2.4),

$$\left. \begin{array}{c} xy + y - x \geq 0 , \\[2mm] x + y - xy \geq 0 , \\[2mm] x - y + xy \geq 0 , \end{array} \right\} \qquad (1.4.24)$$

we have to discuss the stability regions.

(1.4.2.1) Condition $S_1 > 0$.

The condition $S_1 > 0$ is discussed by the transformation (1.3.14) to the cartesian coordinates u, v (1.3.9) :

$$\bar{S}_1(u, v; k) \equiv \frac{5 + 2k}{3}\left(u + \sqrt{3}\,\frac{2 + k}{5 + 2k}\right)^2 - v^2 - \frac{(k + 1)^2 - 2}{5 + 2k} > 0 . \quad (1.4.25)$$

The boundaries of (1.4.25) are conic sections :

a) $-\infty < k < -\dfrac{5}{2}$: circles,

b) $k = -\dfrac{5}{2}$: parabola,

c) $-\dfrac{5}{2} < k < -1 -\sqrt{2}$: hyperbolas with u – vertices,

d) $k = -1 -\sqrt{2}$: 2 straight lines,

e) $-1 -\sqrt{2} < k < -1 +\sqrt{2}$: hyperbolas with v – vertices,

f) $k = -1 +\sqrt{2}$: 2 straight lines,

g) $-1 +\sqrt{2} < k$: hyperbolas with u – vertices.

A detailed discussion of the conic sections shows :

1. $S_1(x,y; k) > 0$ for $k < 0$; instability regions in the
     shape triangle will occur only for $k \geq 0$.

2. $S_1(x,y; k) = S(y,x; k)$: Symmetry with respect to

     $u = 0$ $(B = C)$.

In Fig. 9 the instability regions are shown for different values
of k .

(1.4.2.2)Condition $S_2 > 0$.

This condition is very easy to discuss in MAGNUS' coordinates
because the boundaries are of the two types

(1.4.26) $x = \dfrac{1}{1-k}$ (y arbitrary),

(1.4.27) $y = \dfrac{1}{1-k}$ (x arbitrary).

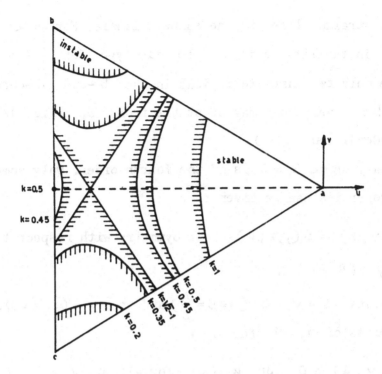

Fig. 9. Condition $S_1 > 0$ for STAUDE's permanent rotations.

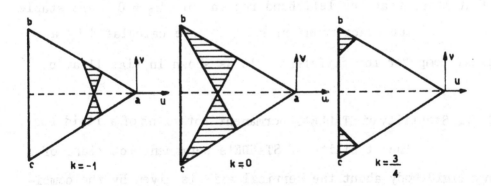

Fig. 10. Condition $S_2 > 0$ for STAUDE's permanent rotations.
The shaded regions are instable.

These are straight lines in the shape triangle. For $-\infty < k < \frac{1}{2}$ there are instability regions of bow–tie–type, for $\frac{1}{2} \leq k \leq 1$ there are only two unstable regions in the b –and c –corners, and for $k > 1$ every body may be stable from S_2 (Fig. 10).

(1.4.2.3)Condition $S_3 \geq 0$.

The boundary curve $S_3 = 0$ is of the fourth order. Only some simple properties can be given :

1. $S_3(x,y;k) = S_3(y,x;k)$: Symmetry with respect to

   $x = y$ $(B = C)$.

2. The points $x = y = 2$ $(B = C = \frac{A}{2})$ and $x = y = 0$ $(A = 0, B = C)$ are points of $S_3 = 0$ for all k .

3. $S_3(x,y;k) \geq 0$ for $k \leq 0$ for all x , y .

4. Related to S_3 the points of the symmetrical axis $B = C$ are stable if $x \geq 4k$ because $S_3(x,x;k) = (x-2)^2 x (x-4k)$.

   That means that the left hand regions of $S_3 = 0$ are stable.

   The exact boundaries   are calculated by a

digital computer for different   's as shown in Fig. 11 a, b.

1.4.2.4.Stability of STAUDE's permanent rotations of a rigid body.

        The stability of STAUDE's permanent rotations of a heavy rigid body about the vertical axis is given by the combinations of the three conditions (1.4.21 – 1.4.23). For $k < 0$ (the centre of mass S lies under the suspension point 0) the stability regions are determined by $S_2 > 0$ and the stability

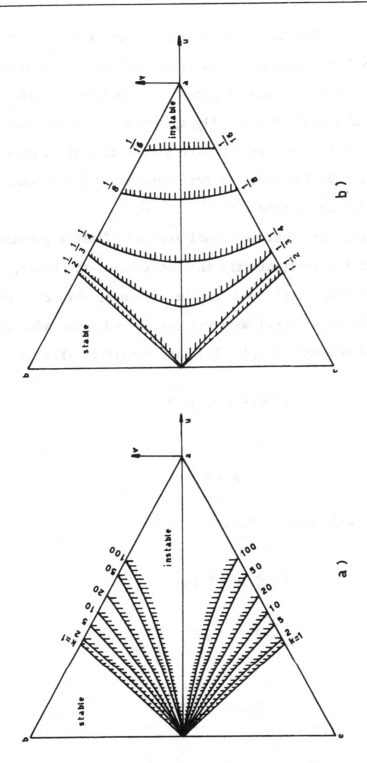

Fig. 11. Condition $S_3 > 0$ for STAUDE's permanent
rotations for different values of k.

card is as shown in Fig. 10 for negative k . For $k \geq \frac{1}{2}$ the
conditions $S_1 > 0$ and $S_3 \geq 0$ are contradictory and there is no
stable region. There are small regions of stability for $0 \leq k \leq \frac{1}{2}$
only (see Fig. 12 a – e). Practically speaking the heavy, unsymmetrical rigid body with a mass centre higher than the suspension
point is stable only for some few configurations of the moments
of inertia in the neigbourhood of $B = C = \frac{A}{2}$.

From the stability conditions of STAUDE's permanent
rotations about the vertical axis the rotation of the heavy,
symmetrical gyroscope with $B = C$ can be examined. This case of
rotations of the heavy rigid body was considered independently
by LAGRANGE and POISSON [2, 3] . The main condition will be

$$S_3(x, x ; k) \geq 0 \quad .$$

Hence

(1.4.28) $x \geq 4k \quad ,$

which means the well known relation

(1.4.29) $A^2 \omega_0^2 \geq 4 B m g s \quad .$

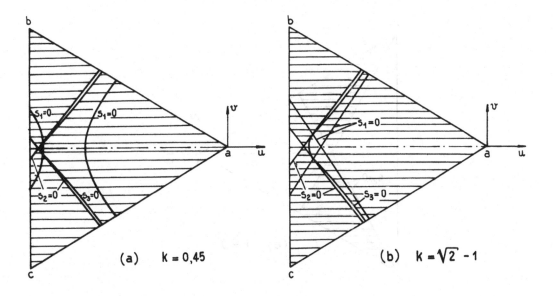

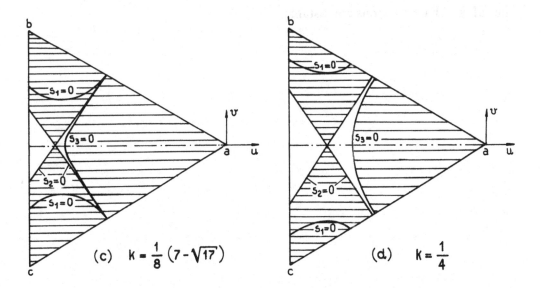

Fig. 12. Stable regions for STAUDE's permanent rotations for different values of k. Shaded regions are instable.

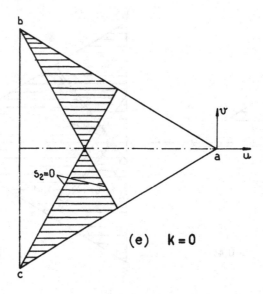

Fig. 12. Stable regions for STAUDE's permanent rotations for different values of k. Shaded regions are instable.

References

[1] W. SCHIEHLEN : Dynamics of Satellites. Texbook, Centre
       International des Sciences Mécaniques (CISM), Udine
       1970.

[2] K. MAGNUS : Gyro-Dynamics. Textbook, Centre International
       des Sciences Mécaniques (CISM), Udine 1970.

[3] K. MAGNUS : Kreisel, Theorie und Anwendungen. Springer-
       Verlag, Berlin-Heidelberg-New York 1971.

[4] K. MAGNUS : Drehbewegungen starrer Körper im zentralen
       Schwerefeld. In : Applied Mechanics, ed. by H.
       GÖRTLER, Springer-Verlag, Berlin-Heidelberg-New
       York 1966, pp. 88-98.

[5] W. SCHIEHLEN und O. KOLBE : Gravitationsstabilisierung
       von Satelliten auf elliptischen Bahnen. Ing.-Arch.
       38 (1969), pp. 389-399.

[6] O. STAUDE : Über permanente Rotationsachsen bei der Bewe-
       gung eines schweren Körpers um einen festen Punkt.
       J. Reine u. Angew. Math. 113 (1894), pp. 318-334.

Chapter 2
Stability of Mechanical Systems

2. 1. Introduction

The stability of discrete mechanical systems can be studied by the well-known method of LJAPUNOV or by geometrical and algebraic criteria in the case of autonomous linear systems (see HAHN [1]). Based on investigations of THOMSON and TAIT [2] additional stability theorems exist for linear autonomous mechanical systems which are much more easy to handle. This lecture deals with two such qualitative stability theorems. For a better understanding we will describe the nature of these prob_lems more exactly.

Mechanical systems of n discrete generalized masses may be often characterized by a linear (or linearized) autonomous matrix differential equation

$$(2.1.1) \qquad M\ddot{x} + (D + G)\dot{x} + (K + N)x = 0 .$$

Here x means the n-dimensional displacement vector from the equilibrium position, M the symmetric positive definite mass matrix, D the symmetric damping matrix of dissipative forces, G the skew-symmetric matrix of gyroscopic forces, K the non-singular symmetric spring-matrix of forces being derivable from a potential, and N is the skew-symmetric matrix of nonconserva-

tive forces depending on the displacements alone. Such mechanical systems occur in the analysis of gyroscopic systems, e.g.
in the analysis of high-speed rotors [3] . Another example comes
from the problems of satellite attitude control. One is often
interested in attitude stability with respect to an orbiting
reference frame. Because this frame is rotating, gyroscopic
forces arise from Coriolis and centrifugal effects. The linearized differential equations of small angle motion about equilibrium are typically of the form (2.1.1)(see [4]).

The stability of the equilibrium $x = 0$ of (2.1.1)
may in many instances be conveniently investigated through the
use of a set of qualitative stability theorems, recently
compiled by MAGNUS (e.g. see [5]). Because of the special structure of (2.1.1) and its matrices there are many state-
ments of stability without knowledge of the eigenvalues of
(2.1.1). The most famous of these stability theorems is one of
THOMSON - TAIT - CHETAEV which was formulated by THOMSON and
TAIT [2] and proved by CHETAEV [6] . It stated that for $N = 0$
and a positive definite D the stability of the solutions of
(2.1.1) is the same as the stability of the solutions of the
abbreviated equation

$$M\ddot{x} + Kx = 0 \ . \qquad\qquad (2.1.2)$$

The concept of Ljapunov stability is thereby used. Because the
equilibrium of (2.1.2) is stable if and only if K is a posi-

tive definite matrix, the equilibrium of (2.1.1) is asymptotical
ly stable by the assumption on D in the stable case of the mentioned theorem. Recently ZAJAC [7, 8] and PRINGLE [9, 10] extend
ed that theorem.

For all these convenient tests of stability there
are some deficiencies. For example, what does occur if the damping matrix is merely positive semidefinite ? As shown in [6] the
THOMSON - TAIT theorem is not true in general. Because in many
practical systems the damping is in fact only positive semidefinite, the requirement of definiteness is very restrictive.
Therefore, in the following part necessary and sufficient conditions are given for the validity of the THOMSON - TAIT -
CHETAEV theorem if D is only positive semi-definite.

Another problem arises from the linearization tech-
nique. The stability behaviour of a nonlinear system is equivalent to that of the equation in the first approximation only
if the linearized system is unstable or asymptotically stable.
In the critical case of oscillatory stability of (2.1.1) a state
ment of the stability behaviour of the nonlinear system is not
possible in general without use of second order terms. However,
for mechanical systems a sufficient test of stability is possible
in such critical cases. That will be discussed in the last section of chapter 2.

2. 2. Asymptotic stability of linear mechanical systems with semidefinite damping

2. 2. 1. Pervasive damping

We would like to extend the above-mentioned THOMSON-TAIT - CHETAEV theorem to positive semidefinite damping matrices D . Therefore, we consider the stability behaviour of

$$M\ddot{x} + (D + G)\dot{x} + Kx = 0 \qquad (2.2.1)$$

with

$$
\left.
\begin{aligned}
M &= M^T > 0 , \\
D &= D^T \geq 0 , \\
G &= -G^T , \\
K &= K^T , \quad \det K \neq 0 .
\end{aligned}
\right\} \qquad (2.2.2)
$$

For that reason we introduce the concept of pervasive damping as given by PRINGLE [9] . The damping forces $-D\dot{x}$ are called pervasive if and only if the quadratic form

$$P = -\dot{x}^T(t)D\dot{x}(t) \qquad (2.2.3)$$

is nonpositive and vanishes for all $t > 0$ if and only if $x(t) \equiv 0$. Hence (2.2.1) is pervasively damped if there is no motion $x(t) \neq 0$ with $P \equiv 0$.

The problem is therefore, when is (2.2.1) pervasively damped ? According to (2.2.3) this problem can be decided
only by the knowledge of the solution of (2.2.1). We are not,
however, interested in this concept, since for that we have to
know the eigenvalues of (2.2.1) and hence the stability behaviour.
Therefore, we have to look for a simple criterion of pervasiveness.
Certain possibilities exist. ROBERSON [11] solves the problem by
means of graph theory and CONNELL [12] uses solutions of suitable
eigenvalues problems. Both methods are cumbersome and from the
point of view of structural analysis, unsatisfactory. Therefore,
in [13] a new method of testing pervasiveness is given by the author.

If a motion $\bar{x}(t) \neq 0$ with the property $P = 0$
exists then this motion can be characterized by the state
space vector

$$(2.2.4) \qquad \bar{y}(t) = \begin{bmatrix} K & 0 \\ 0 & M \end{bmatrix}\begin{bmatrix} \bar{x}(t) \\ \dot{\bar{x}}(t) \end{bmatrix}$$

Because of $\quad D\dot{\bar{x}}(t) = 0 \qquad$ we have

$$(2.2.5) \quad \dot{\bar{y}}^T(t) = -\bar{y}^T(t) = \begin{bmatrix} 0 & I \\ -M^{-1}K & -M^{-1}G \end{bmatrix} = -\bar{y}^T(t)A ,$$

$$(2.2.6) \quad \bar{y}^T(t)\begin{bmatrix} 0 \\ M^{-1}D \end{bmatrix} = \bar{y}^T(t)B = \dot{\bar{x}}^T(t)D = 0 .$$

The derivation of (2.2.6) with respect to time t leads with

regard to (2.2.5) to

$$\frac{d^k}{dt^k}(\bar{y}^T(t)B) = (-1)^k \bar{y}(t)A^k B \equiv 0, \quad (k = 0, 1, \ldots). \quad (2.2.7)$$

As a result of CAYLEY-HAMILTON theorem the power k of A^k is interesting for $k = 0, 1, \ldots, 2n-1$. The relation (2.2.7) effects

$$\text{rank } Q < 2n \quad , \quad\quad\quad (2.2.8)$$

where

$$Q = [B \vdots AB \vdots \ldots \vdots A^{2n-1}B] . \quad\quad (2.2.9)$$

Therefore, if (2.2.1) is not pervasively damped then relation (2.2.8) is true. In [13] it is also proved that

$$\text{rank } Q = 2n \quad\quad\quad (2.2.10)$$

is equivalent to pervasive damping.

In modern control theory the matrix Q plays an important part and is called the controllability matrix of the controlled state space system

$$\dot{z} = Az + Bu \quad\quad\quad (2.2.11)$$

[14] . Hence we can say : The mechanical system (2.2.1) with matrices (2.2.2) is pervasively damped if and only if the controlled system

$$M\ddot{x} + G\dot{x} + Kx = Du \quad\quad (2.2.12)$$

with control u is completely controllable.

2. 2. 2. Stability

To check the stability behaviour of the mechanical system (2.2.1, 2.2.2), state space notation suggests itself :

(2.2.13)
$$y = \begin{bmatrix} x \\ \dot{x} \end{bmatrix} \quad ,$$

(2.2.14)
$$\dot{y} = \left[\begin{array}{c|c} 0 & I \\ \hline -M^{-1}K & -M^{-1}(D+G) \end{array} \right] y \quad .$$

The HAMILTON–function of (2.2.1) reads

(2.2.15)
$$H = \frac{1}{2} y^T \left[\begin{array}{c|c} K & 0 \\ \hline 0 & M \end{array} \right] y$$

and its derivative is as follows

(2.2.16)
$$\dot{H} = y^T \left[\begin{array}{c|c} 0 & 0 \\ \hline 0 & -D \end{array} \right] y = P \quad .$$

By (2.2.16) the importance of pervasive damping is obvious. By the application of the HAMILTON–function as a LJAPUNOV–function we obtain the desired extension of the THOMSON – TAIT theorem to semidefinite damping matrices.

Theorem : If the mechanical system (2.2.1) with the matrices
                (2.2.2) is pervasively damped, i.e. if relation (2.2.10)

holds, then the stability behaviour of (2.2.1) is the
same as the stability behaviour of the abbreviated system (2.1.2). More exactly, the number of unstable modes of (2.2.1) is equal to the number of negative eigenvalues of K. Hence, the equilibrium $x = 0$ of
(2.2.1) is asymptotically stable if and only if K is
positive definite and (2.2.10) holds.

The proof is based on the stability and instability
theorems of LJAPUNOV and CHETAEV [1] . In the case of instability the dimension of the region of unstable initial conditions is
equal to the number of negative eigenvalues of K and, therefore,
equal to the number of unstable modes (detailed discussion see
[7, 8]).

2. 2. 3. Example

We follow the example of CONNELL [12] . The linearized equations of rotational motion are considered for a hinged,
two-body, five-degree-of-freedom gravity-gradient satellite in a
circular orbit and with the **centres** of mass of each body at the
hinge point. These equations are of the same form as (2.2.1) with
the pitch equations, uncoupled from the roll-yaw equations, so
that each set may be considered separately. With J_α as the principal moments of inertia of the composite system ($\alpha = 1$ nominal
roll, $\alpha = 2$ nominal pitch, $\alpha = 3$ nominal yaw), I_α^j as the principal moments of inertia of the main body ($j = 1$) and the auxi-

liary body ($j = 2$), d_i ($i = 1, 2$) as two damping constants divided by the orbital frequency, and K_i ($i = 1, 2$) as two spring constants divided by the square of the orbital frequency, the matrices for the pitch equation are

(2.2.17)
$$M_1 = \begin{bmatrix} J_2 & I_2^2 \\ I_2^2 & I_2^2 \end{bmatrix} \quad , \qquad D_1 = \begin{bmatrix} 0 & 0 \\ 0 & d_1 \end{bmatrix} \, ,$$

$$G_1 = 0 \, , \qquad K = \begin{bmatrix} 3(J_1 - J_3) & 3(I_1^2 - I_3^2) \\ 3(I_1^2 - I_3^2) & 3(I_1^2 - I_3^2) + K_1 \end{bmatrix} \, ,$$

and the matrices for the roll–yaw equations are

(2.2.18)
$$M_2 = \begin{bmatrix} J_1 & 0 & I_1^2 \\ 0 & J_3 & 0 \\ I_1^2 & 0 & I_1^2 \end{bmatrix} \, ,$$

$$G_2 = \begin{bmatrix} 0 & J_1 + J_3 - J_2 & 0 \\ J_2 - J_1 - J_3 & 0 & I_2^2 - I_1^2 - I_3^2 \\ 0 & I_1^2 + I_3^2 - I_2^2 & 0 \end{bmatrix} \, ,$$

$$D_2 = \begin{bmatrix} 0 & 0 & 0 \\ 0 & 0 & 0 \\ 0 & 0 & d_2 \end{bmatrix} \, ,$$

$$K_2 = \begin{bmatrix} 4(J_2 - J_3) & 0 & 4(I_2^2 - I_3^2) \\ 0 & J_2 - J_1 & 0 \\ 4(I_2^2 - I_3^2) & 0 & 4(I_2^2 - I_3^2) + K_2 \end{bmatrix} \, .$$

Firstly, we will discuss the stability conditions of the pitch motion. Because of $\lambda_2 - I_2^2 = I_2^1 > 0$ the mass matrix M_1 is positive definite. The matrix K_1 is pos. definite provided

$$\lambda_1 - \lambda_3 > 0 \qquad (2.2.19)$$

and

$$\frac{K_1}{I_2^1} > \frac{-3\beta_1\beta_2}{(I_2^1/I_2^2)\beta_1 + \beta_2} , \qquad (2.2.20)$$

where

$$\beta_j = (I_1^j - I_3^j)/I_2^j \qquad (j = 1, 2) . \quad (2.2.21)$$

Because of the pervasiveness of the damping, we have to test relation (2.2.10) which reads here as

$$\text{rank } Q_1 = \text{rank} \begin{bmatrix} 0 & M_1^{-1}D_1 & 0 & -M_1^{-1}K_1M_1^{-1}D_1 \\ M_1^{-1}D_1 & 0 & -M_1^{-1}K_1M_1^{-1}D_1 & 0 \end{bmatrix} \qquad (2.2.22)$$
$$\overset{(?)}{=} 4 .$$

It is obvious that rank $Q_1 = 4$ if and only if

$$\text{rank} \begin{bmatrix} D_1 & \vdots & K_1M_1^{-1}D_1 \end{bmatrix} = 2 \qquad (2.2.23)$$

which is equivalent to

$$(\lambda I + K_1 M_1^{-1}) \begin{bmatrix} 0 \\ 1 \end{bmatrix} \neq 0 \quad \text{for all real } \lambda \qquad (2.2.24)$$

or

$$K_1 \begin{bmatrix} -I_2^2 \\ d_2 \end{bmatrix} \neq \frac{3 I_2^1 (I_1^2 - I_3^2) + k_1 d_2}{\det M_1} \begin{bmatrix} 0 \\ 1 \end{bmatrix} .$$

Hence, relation (2.2.24) is true for

(2.2.25) $\beta_1 \neq \beta_2$.

The damping is pervasive and, therefore, the equilibrium position of pitch motion is asymptotically stable if (2.2.19) and
(2.2.20) hold.

         The mass matrix M_2 of the roll-yaw motion is posi-
tive definite. Also the spring matrix is positive definite if

(2.2.26) $d_2 - d_3 > 0$,

(2.2.27) $d_2 - d_1 > 0$,

and

(2.2.28) $\dfrac{K_2}{I_1^1} > \dfrac{-4 \, \delta_1 \, \delta_2}{(I_1^1/I_1^1) \delta_1 + \delta_2}$,

where

(2.2.29) $\delta_j = (I_2^j - I_3^j)/I_1^j$ $(j = 1, 2)$.

For the investigation of (2.2.10) we remember that relation
(2.2.10) is equivalent to

(2.2.30) $\text{rank} \begin{bmatrix} A - \lambda I, & B \end{bmatrix} = 2n$, $\lambda \in \sigma (A)$

(see [15]) where $\sigma(A)$ is the spectrum of A. Here condition (2.2.30) reads

$$\text{rank} \left[\begin{array}{c|c} -\lambda M_2 & M_2 \\ \hline -K_2 & -G_2-\lambda M_2 \end{array} \right], \; d_2 \begin{bmatrix} 0 \\ 0 \\ 0 \\ 0 \\ 0 \\ 1 \end{bmatrix} = 6 \;. \quad (2.2.31)$$

A short manipulation of (2.2.31) leads to ($d_2 > 0$ assumed)

$$\text{rank} \left[\lambda^2 M_2 + \lambda G_2 + K_2 \right], \; \begin{bmatrix} 0 \\ 0 \\ 1 \end{bmatrix} = 3 \;, \quad (2.2.32)$$

which is true by (2.2.26 – 28), because the sufficient condition

$$\left[\lambda^2 M_2 + \lambda G_2 + K_2 \right] \begin{bmatrix} 0 \\ 0 \\ 1 \end{bmatrix} \neq 0 \quad (2.2.33)$$

is satisfied for all real satellites.

Therefore, if the requirements, given by (2.2.26 – 28), are satisfied the roll-yaw motions are asymptotically stable. In conjunction with the conditions (2.2.19 – 20) the small angle motions of the two-body-satellite about equilibrium are asymptotically stable.

2. 3. Statement of Stability Behaviour of Nonlinear Mechanical Systems by the Method of First Approximation

From LJAPUNOV's stability theory [1] it is known that the stability behaviour of an autonomous nonlinear me-

chanical system is equivalent to that of the linearized equation
of type (2.1.1) if the equilibrium position $x = 0$ of (2.1.1) is
either asymptotically stable or unstable. In the critical case
of oscillatory stability no statement is possible generally speaking. However, as mentioned in the introduction, also in that
case for mechanical systems a simple method exists to check the
stability of the nonlinear system by the linearized model.

Following POPP [16, 17] the linearized model of the
nonlinear mechanical system may be given by (2.2.1),

$$(2.3.1) \qquad M\ddot{x} + (D + G)\dot{x} + Kx = 0 ,$$

with matrices

$$(2.3.2) \quad M = M^T > 0 , \quad D = D^T , \quad G = -G^T , \quad K = K^T .$$

Because of the critical case we assume that (2.3.1) is oscillatory stable, i.e. at least one mode of (2.3.1) is oscillatory
stable while the remaining modes are asymptotically stable. Now,
what is there to say about the stability behaviour of the nonlinear system, if we know that the total energy of the mechanical problem does not increase (or : increases only) ?

Theorem : If the total derivative with respect to time t of the
HAMILTON-function of an autonomous nonlinear mechanical system is negative semidefinite or identically
vanishing (or positive definite) in a neighbourhood
of the equilibrium, and if the linearized model is

given by (2.3.1) and (2.3.2), then the equilibrium
$x = 0$ of the nonlinear system is stable (or : unstable) if

$$K > 0. \qquad\qquad (2.3.3)$$

Physically the theorem means that the nonlinear system is stable (or : unstable) if the equilibrium is statically
stable and the total energy does not increase (or : increases
only). The proof of the theorem runs by application of the HA-
MILTON-function as LJAPUNOV-function of the nonlinear system
and its approximation by the linearized one. A detailed proof
is given in [17] .

As an example of the theorem we look for the stability of gravity-gradient stabilized satellites on circular orbits. As shown in [4] the linearized model is oscillatory stable
if the conditions (1.1.1) and (1.1.2) of chapter 1 are satisfied. But only in the case of relation (1.1.1) the satellite is
statically stable. Therefore, the configurations of the satellite, being placed in the so-called LAGRANGE region (1.1.1),
are stable in the nonlinear case, too. Satellites, characterized by (1.1.2) (DELP region), are statically unstable and are
stabilized only by gyroscopic forces ; the theorem does not
yield a statement of stability behaviour in the nonlinear consideration. That may be a suggestion that the moments of inertia of a
gravity-gradient satellite should be chosen in the LAGRANGE

region (similar suggestions are found in [4]).

References

[1] W. HAHN : Stability of Motion. Springer-Verlag, Berlin-
          Heidelberg – New York 1967.

[2] W. THOMSON and P.G. TAIT : Treatise on Natural Philoso-
          phy, Part I, Cambridge 1921.

[3] G. SCHWEITZER : Critical Speeds of Gyroscopes. Textbook,
          Centre International des Sciences Mécaniques (CISM),
          Udine 1970.

[4] W. SCHIEHLEN : Dynamics of Satellites. Textbook, Centre
          International des Sciences Mécaniques (CISM), Udine
          1970.

[5] K. MAGNUS : Gyro-Dynamics. Textbook, Centre International
          des Sciences Mécaniques (CISM), Udine 1970.

[6] N.G. CHETAEV : The Stability of Motion (Translation from
          the Russian by M. NADLER.) Pergamon Press, New York-
          Oxford – London – Paris 1961.

[7] E.E. ZAJAC : The Kelvin – Tait – Chetaev Theorem and Ex-
          tensions. J. Astronaut. Sci. 11 (1964), No. 2, pp.
          46-49.

[8] E.E. ZAJAC : Comments on "Stability of Damped Mechanical
          Systems" and a Further Extension. J. AIAA 3 (1965)
          pp. 1749 – 1750.

[9] R. PRINGLE, JR : Stability of Damped Mechanical Systems.
          J. AIAA 3 (1965) p. 363.

[10] R. PRINGLE, JR : On the Stability of a Body with Connec-
          ted Moving Parts. J. AIAA 4 (1966) pp. 1395-1404.

[11]   R.E. ROBERSON : Notes on the Thomson – Tait – Chetaev
        Stability Theorem. J. Astronaut. Sci. 15 (1968)
        pp. 319 – 322.

[12]   G.M. CONNELL : Asymptotic Stability of Second-order
        Linear Systems with Semidefinite Damping. J. AIAA 7
        (1969) pp. 1185 – 1187.

[13]   P.C. MÜLLER : Asymptotische Stabilität von linearen
        mechanischen Systemen mit positiv semidefiniter
        Dämpfungsmatrix.Z. Angew. Math. Mech. 51 (1971), pp.
        T 197 – T 198.

[14]   R.E. KALMAN, Y.C. HO, and K. NARENDRA : Controllability
        of Linear Dynamical Systems. Contr. Diff. Equations
        1 (1963) pp. 189 – 213.

[15]   M.L.J. HAUTUS : Controllability and Observability Condi-
        tions of Linear Autonomous Systems. Nederl. Akad.
        Wet., Proc., Ser. A 72 (1969) pp. 443 – 448.

[16]   K. POPP : Stabilitätsuntersuchung an Zweikörper-Satelli-
        ten.Z.Angew.Math.Mech.51(1971),pp.T 205–T 207.

[17]   K. POPP : Stabilität mechanischer Systeme mit nicht
        durchdringender Dämpfung. To be published in
        Z. Angew. Math. Phys. 1972.

Chapter 3
Error Analysis of Inertial Platforms

3. 1. Introduction

Error analysis plays an important part in the design and development of inertial navigation systems. It is
through error analysis that the requirements on instruments
accuracies are specified and the system performance is evalu-
ated. Errors in an inertial navigation system can arise in a
number of ways, i.e. accelerometer and gyro inaccuracies, misalignment of the instruments on the platform, initial errors in
in the platform alignment, computational errors, and approximations in the mechanization of the system equations. In this
chapter the error equations for arbitrary inertial navigation
systems will be developed.

In the time propagation of the position errors for
an inertial guidance system, it is essential to consider the
platform misorientations. By dealing with the platform misorien-
tation with respect to the indicated position instead of the
true position it is possible to develop the platform error
equations which are independent of the position errors. Therefore, after a brief description of inertial platforms a derivation of the platform error equation is given. Following this,
position error equations will be derived for which the platform

errors are the driving functions. Error progation will be dis-

cussed at the end of this chapter.

This chapter is based on references [1, 2, 3] .

3. 2. Inertial platforms

The complete inertial navigation system is composed

of a group of subsystems. The platform, platform electronics,

position computer, and power supply are always included. Opti-

cal systems are mounted, and the star-pointing and error angle

positions of the computer are included only when stellar moni-

toring of platform performance is to be carried out. When the

navigator is not designed to be self-checked and self-aligned,

two additional subsystems are required. These are the ground

checkout equipment used to determine that navigator subsystems

are operating correctly and the alignment equipment used to set

initial conditions for the stable platform.

Let us now examine the subsystem most closely iden-

tified with inertial navigation : the stable platform.

The task of the stable platform is to maintain ac-

celeration-sensing instruments in a precise known angular orien

tation in space. The stable platform is composed of a mount for

gyroscopes and acceleration-measuring devices, known as the

stable element,the inertial instruments themselves, and a gim-

baling system which permits the required vehicle angular motion

about the stable element.

The stable-element structure is designed to provide
mounting surfaces for the inertial instruments with accurate
known angular relations between them. The accuracy with which
these relative angles must be maintained varies according to
the application and mechanization of the inertial navigator,
but for the most critical elements it will be of the order of
10 seconds of arc. These relative orientations must be maintained nor with standing changes of vibration, temperature
cycling, acceleration, or other environmental conditions en-
countered during the application of the system.

Gyroscopes are mounted on the stable element to
act as dip sensors about each of three mutually orthogonal axes
(stable-element coordinates axes), and in some cases they provide an appreciable part of the stabilizing torque themselves
over a part of the frequency range of disturbing torques. Stabilization is obtained mainly through the use of dip sensor error
signals to control platform torquers. Electromagnetic torques,
however, may be applied to the gyroscopes themselves in order
to cause them to precess at a known rate. This is done to maintain the stable-element coordinate system in some known rela-
tionship to a reference coordinate system.

The stable element requires platform servo-loops
to assist the gyroscopes in stabilizing the reference element.
Indeed, in modern design the trend is towards the use of gyroscopes mainly as disturbance-torque sensors than as counter-

torque producers. Servos must maintain the platform and gyroscope angles within specified limits in spite of disturbances
imposed on the platform. These disturbances may enter the servo-
loop as torques or angular motions.

The accelerometers sense components of vehicle acceleration in the coordinates of the stable element. From these
accelerations, changes in velocity and position of the vehicle
are determined. From NEWTON's law the acceleration measurements
are given by

$$b = \ddot{R} + \dot{\omega} \times R + 2\omega \times \dot{R} + \omega \times (\omega \times R) + g \qquad (3.2.1)$$

where b means the acceleration vector in the platform coordinates, R the position vector of the vehicle from the **centre of**
earth, ω the angular-velocity vector of the stable element,
and g the gravity vector. Because of the angular-velocity of
the inertial platform,EULER,CORIOLIS-and centrifugal accelerations have to be considered in $(3.2.1)$.By integration of $(3.2.1)$
the angular-velocity ω and the position vector R are determined.

Because of the mechanization of $(3.2.1)$, or more
exactly,the translation of measured accelerations into veloci-
ties and positions, we have to introduce some coordinate systems. The reference system is defined as an assumed coordinate
system $\{x_i\}$ which has the origin in the centre of the earth
and which rotates with ω . The **coordinate** system $\{x_i^p\}$ of the

stable platform has to correspond to $\{x_i\}$, i.e. the axes of
the coordinate system are parallel but the origin of $\{x_i^p\}$ is
in agreement with the position of the vehicle. Equally, the computer coordinate system $\{x_i^c\}$, in which the calculation of ω
and R is mechanized, has to be in agreement with $\{x_i\}$. The
choice of the coordinate system $\{x_i\}$ for a specific system mechanization depends on many considerations. For example, in many
navigational systems latitude φ and longitude λ are the desired outputs, and consequently the system should be mechanized
to yield these outputs directly. The latitude-longitude coordinate system $\{x_i^s\}$ is characterized by a coordinate plane which
is locally level. The three axes are pointing north, west, and
vertical as shown in Fig. 1. The reference angular-velocity
yields in

$$(3.2.2) \quad \begin{cases} \omega_1^s = \omega^E \cos \varphi - \dfrac{v_2^s}{R} \, , \\[2ex] \omega_2^s = \dfrac{v_1^s}{R} \, , \\[2ex] \omega_3^s = \omega^E \sin \varphi - \dfrac{v_2^s}{R} \tan \varphi \, , \end{cases}$$

where v_i^s are the components of the vehicle velocity, ω^E the
earth rate and R the distance between the vehicle and the
centre of earth.

For illustration a diagram of a stable platform
with three accelerometers and three rate gyros is presented in

Fig. 2. The stabilization loops are omitted.

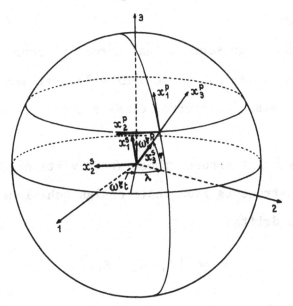

Fig. 1. Latitude-longitude coordinate system $\{x_i^s\}$ and platform coordinate system $\{x_i^p\}$.

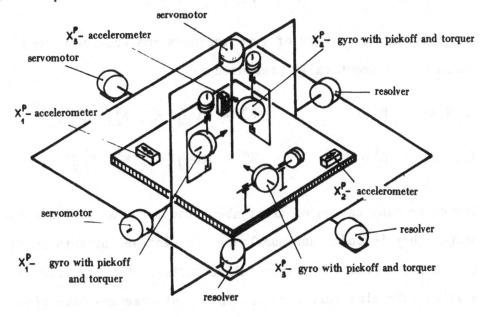

Fig. 2. Diagram of a stable platform.

3. 3. Error sources

The error sources that normally occur in an inertial navigation system can be conveniently assigned to one of
three basic classes. These error classes are :

1. Physical—component errors. These are deviations of the physical inertial instruments from their design behaviour. The main
errors are gyro drifts

$$(3.3.1) \qquad\qquad \varepsilon = [\varepsilon_1 \quad \varepsilon_2 \quad \varepsilon_3]^T ,$$

accelerometer bias

$$(3.3.2) \qquad\qquad \nabla = [\nabla_1 \quad \nabla_2 \quad \nabla_3]^T ,$$

and scale factor errors of gyro torquers and accelerometers
which are mathematically presented as

$$(3.3.3) \qquad K^G \omega \quad , \qquad\qquad K^G = \text{diag}[K_1^G, K_2^G, K_3^G] ,$$

$$(3.3.4) \qquad K^A b \quad , \qquad\qquad K^A = \text{diag}[K_1^A, K_2^A, K_3^A] .$$

There are many causes of drift about the input axis of a gyroscope. They include mass unbalance, convection currents in flo-
tation fluid from temperature differentials, bearing friction,
bearing anisoelasticity, etc.. The accelerometers have bias
errors because of instability of the biasing equipment.

2. Construction errors. These are errors in the basic overall
system construction such as mechanical alignment errors of components on the platform. In the linearized case the influence
of the nonorthogonality of the gyro input axes does not yield
the angular velocity ω but

$$\omega^6 = (I + \tilde{\psi})\omega \quad , \tag{3.3.5}$$

where I is the unit matrix, and

$$\tilde{\psi} = \begin{bmatrix} 0 & \psi_1 & -\psi_2 \\ -\psi_3 & 0 & \psi_4 \\ \psi_5 & -\psi_6 & 0 \end{bmatrix} \tag{3.3.6}$$

characterizes the misalignment angles of the gyro input axes
with respect to $\{x_i^P\}$ (see Fig. 3). Equally, the acceleration
measurements will be

$$b^A = (I + \tilde{\Gamma})b \tag{3.3.7}$$

with the matrix

$$\tilde{\Gamma} = \begin{bmatrix} 0 & \Gamma_1 & -\Gamma_2 \\ \Gamma_3 & 0 & \Gamma_4 \\ \Gamma_5 & -\Gamma_6 & 0 \end{bmatrix} \tag{3.3.8}$$

of the deviation angles Γ_i ($i = 1,\ldots,6$) of the accelerometers
with respect to $\{x_i^P\}$.

3. Initial conditions. These errors arise from the imperfect
determination of the initial position vector,

$$(3.3.9) \qquad\qquad \Delta R_0 \ ,$$

of the initial velocity vector,

$$(3.3.10) \qquad\qquad \Delta \dot{R}_0 \ ,$$

and of the initial misalignment

$$(3.3.11) \qquad\qquad \alpha_0 \ = \ [\alpha_{10} \ \ \alpha_{20} \ \ \alpha_{30}]^T$$

of the inertial platform, i.e. the misalignment of $\{x_i^P\}$ with
respect to $\{x_i\}$.

3. 4. Platform error equation

In deriving the system error equations it is neces-
sary to consider the three sets $\{x_i\}$, $\{x_i^P\}$, and $\{x_i^C\}$ of coordinate axes as defined in section 2. The angular rates of these
three coordinate sets with respect to inertial space are ω ,
ω^P , and ω^C , respectively. If the angular deviations between
these three sets of coordinates are small, these angular deviations can be represented by a set of angles such that α is the
angle vector between the computer axes and the platform axes, β
is the angle vector between the true axes and the platform axes,
and γ is the angle vector between the true axes and the computer axes. The coordinate transformation matrices which rotate a

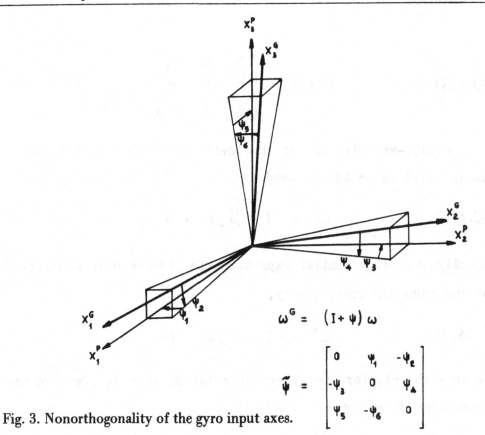

$$\omega^G = (I + \psi)\,\omega$$

$$\tilde{\psi} = \begin{bmatrix} 0 & \psi_1 & -\psi_2 \\ -\psi_3 & 0 & \psi_4 \\ \psi_5 & -\psi_6 & 0 \end{bmatrix}$$

Fig. 3. Nonorthogonality of the gyro input axes.

vector through minus the three error angles may be written

$$I + \tilde{\alpha} = \begin{bmatrix} 1 & \alpha_3 & -\alpha_2 \\ -\alpha_3 & 1 & \alpha_1 \\ \alpha_2 & -\alpha_1 & 1 \end{bmatrix}, \qquad (3.4.1)$$

$$I + \tilde{\beta} = \begin{bmatrix} 1 & \beta_3 & -\beta_2 \\ -\beta_3 & 1 & \beta_1 \\ \beta_2 & -\beta_1 & 1 \end{bmatrix}, \qquad (3.4.2)$$

$$(3.4.3) \qquad I + \tilde{\gamma} = \begin{bmatrix} 1 & \gamma_3 & -\gamma_2 \\ -\gamma_3 & 1 & \gamma_1 \\ \gamma_2 & -\gamma_1 & 1 \end{bmatrix} .$$

The angular-velocity ω^p of the inertial platform with respect to inertial space is obtained by

$$(3.4.4) \qquad \omega^p = (I + \tilde{\beta})\omega + \dot{\beta} .$$

Equally, we get a similar expression for the angular velocity of the computer axes, namely,

$$(3.4.5) \qquad \omega^c = (I + \tilde{\gamma})\omega + \dot{\gamma} .$$

Because the platform angle errors coincide with α, and α becomes from β and γ as

$$(3.4.6) \qquad \alpha = \beta - \gamma ,$$

we get by (3.4.2) and (3.4.3)

$$(3.4.7) \qquad \omega^p = (I + \tilde{\alpha})\omega^c + \dot{\alpha} .$$

Since the angles are small quantities, equation (3.4.7) can be approximated by

$$(3.4.8) \qquad \omega^p = \omega^c + \tilde{\alpha}\omega + \dot{\alpha} ,$$

which is equivalent to the platform error equation

$$(3.4.9) \qquad \dot{\alpha} + \omega \times \alpha = \upsilon ,$$

where

$$v = \omega^P - \omega^C . \qquad (3.4.10)$$

The difference v contains all the errors of the gyroscopes : the scale factor matrix K^G of the gyro torquers, the gyro drift rate ε and the nonorthogonality of the gyro input axes :

$$v = \varepsilon + \tilde{\psi}\omega + K^G\omega . \qquad (3.4.11)$$

Finally, the platform error equation reads

$$\dot{\alpha} + \omega \times \alpha = \varepsilon + \tilde{\psi}\omega + K^G\omega . \qquad (3.4.12)$$

This is a very significant equation which describes the time behaviour of the error angles between the platform and computer axes. It should be noted that the equation is independent of the position error. The angle α , as we shall see in the next section, is a driving function of the position error equation.

3. 5. Position error equation

The accelerometers sense the components of vehicle acceleration in the coordinates $\{x_i^P\}$ of the platform. From these measurements the position of the vehicle is determined by the computer. The statement of the position is given in the coordinates $\{x_i^C\}$ of the computer. Therefore, the position error corresponds to the deviation angles γ . It is caused by the imperfect determination of the acceleration vector b . As

shown in section 3.3 the acceleration error is caused by accelerometer bias (3.3.2), scale factor errors (3.3.4), nonorthogonality of the accelerometer input axes (3.3.7).In addition to these
imperfections errors due to platform misalignment β and to
scale error factors in the compensation loop of the so-called
EULER- accelerations have to be considered. The sum total of all
these terms leads to an acceleration measurement error

(3.5.1) $\Delta b = \nabla + K^{\wedge}b + \tilde{\Gamma}b + b \times \beta + K^{\circ}b^{\circ}$,

where $K^{\circ}b^{\circ}$ is the compensation term :

(3.5.2)
$$\begin{cases} b^{\circ} = b - \dot{\omega} \times R \quad , \\ K^{\circ} = \text{diag} \left[K^{\circ}_1 , K^{\circ}_2 , K^{\circ}_3 \right] . \end{cases}$$

Defining

(3.5.3) $d = \nabla + K^{\wedge}b + \tilde{\Gamma}b + b \times \alpha + K^{\circ}b^{\circ}$,

(3.5.1) reads

(3.5.4) $\Delta b = d + b \times \gamma$.

We now have to look for the acceleration error by means of the
computational mechanization of (3.2.1). The error Δb effects
the deviation between $\{x_i\}$ and $\{x_i^c\}$. The acceleration of
the vehicle is given by (3.2.1) in the reference coordinate sys
tem $\{x_i\}$. Equally, the acceleration vector in the coordinates

of the computer system $\{x_i^c\}$ reads

$$b^c = \ddot{R}^c + \dot{\omega}^c \times R^c + 2\omega^c \times \dot{R}^c + \omega^c \times (\omega^c \times R^c) + g^c \ . \quad (3.5.5)$$

With the abbreviations

$$\Delta\omega = \omega - \omega^c \ , \quad\quad\quad\quad (3.5.6)$$

$$\Delta R = R - R^c \ , \quad\quad\quad\quad (3.5.7)$$

$$\Delta g = g - g^c \ , \quad\quad\quad\quad (3.5.8)$$

another expression is obtained for

$$\Delta b = b - b^c = \Delta\ddot{R} + \Delta\dot{\omega} \times R + \dot{\omega} \times \Delta R + 2\Delta\omega \times \dot{R} +$$

$$+ 2\omega \times \Delta\dot{R} + \Delta\omega \times (\omega \times R) + \omega \times (\Delta\omega \times R) +$$

$$+ \omega \times (\omega \times \Delta R) + \Delta g \ , \quad\quad\quad (3.5.9)$$

where terms of second order in $\Delta\omega$ and ΔR are omitted. With respect to (3.4.5), i.e.

$$\Delta\omega = \dot{\gamma} + \omega \times \gamma \ , \quad\quad\quad\quad (3.5.10)$$

$\Delta\dot{\omega}$ yields

$$\Delta\dot{\omega} = \ddot{\gamma} + \dot{\omega} \times \gamma + 2\omega \times \dot{\gamma} + \omega \times (\omega \times \gamma) \ . \quad (3.5.11)$$

The position error equation is obtained by regarding (3.5.4), (3.5.9) – (3.5.11), and (3.2.1) :

$$[\ddot{\gamma} + 2\omega \times \dot{\gamma} + \dot{\omega} \times \gamma + \omega \times (\omega \times \gamma)] \times R +$$

$$+ 2(\dot{\gamma} + \omega \times \gamma) \times \dot{R} + (\ddot{\gamma} + \omega \times \gamma) \times (\omega \times R) +$$

$$(3.5.12) \quad + \omega \times [(\dot{\gamma} + \omega \times \gamma) \times R] - [\ddot{R} + \dot{\omega} \times R + 2\omega \times \dot{R} + \omega \times (\omega \times R) + g] \times \gamma +$$

$$+ \Delta \ddot{R} + 2\omega \times \Delta \dot{R} + \dot{\omega} \times \Delta R + \omega \times (\omega \times \Delta R) + \Delta g = d \quad .$$

This position error equation is true for all inertial navigation systems. In the case of reference coordinate systems $\{x_i\}$
rotating merely with constant angular velocity ω the position
error equation is simplified to

$$(3.5.13) \quad \Delta \ddot{R} + 2\omega \times \Delta \dot{R} + \omega \times (\omega \times \Delta R) + \Delta g = d \quad .$$

The driving function d (3.5.3) depends on the platform error
as mentioned in section 3.4.

In the case of time-varying ω the position error
equation (3.5.12) holds. Because the angular velocity often shows
only a slight variation, i.e. $\dot{\omega} \approx 0$, the truncated equation
(3.5.13) is used in such cases, too. Therefore, the propagation
of physical-components errors, construction errors and initial
misalignments will be discussed in the next section using
(3.5.13).

3. 6. Error propagation

The determination of the position error propagation requires the solution of the platform equation (3.4.12) as

a first step. The solution furnishes the driving function for
the position error equation (3.5.13), which can then be solved.
For that we assume constant angular velocity ω and constant errors. Thus, the time-varying stochastic errors as gyro drifts
and accelerometer bias are approximated by their mean values.

By means of linear differential system theory the
solution of (3.4.12) is obtained as

$$\alpha(t) = \frac{\omega^T \alpha_0}{\omega^2}\omega - \frac{1}{\omega}\sin\omega t \cdot \tilde{\omega}\alpha_0 + \cos\omega t \left(\alpha_0 - \frac{\omega^T \alpha_0}{\omega^2}\omega\right) +$$

$$+ \left[\left(I + \frac{\tilde{\omega}^2}{\omega^2}\right)t - \frac{\tilde{\omega}}{\omega^2}(1 - \cos\omega t) - \frac{\tilde{\omega}^2}{\omega^2}\sin\omega t\right] \cdot$$

$$\cdot (\varepsilon + \tilde{\psi}\omega + K^6\omega) ,$$

(3.6.1)

where α_0 is the initial condition (3.3.11) of the platform misalignment, $\omega = \sqrt{\omega^T\omega}$, and

$$\tilde{\omega} = \begin{bmatrix} 0 & -\omega_3 & \omega_2 \\ \omega_3 & 0 & -\omega_1 \\ -\omega_2 & \omega_1 & 0 \end{bmatrix} .$$

(3.6.2)

The most critical part of the misalignment $\alpha(t)$ is the time
increasing term

$$t\left(I + \frac{\tilde{\omega}^2}{\omega^2}\right)(\varepsilon + \tilde{\psi}\omega + K^6\omega) = \omega^T(\varepsilon + \tilde{\psi}\omega + K^6\omega)\frac{t}{\omega^2}\omega .$$

(3.6.3)

The error increases with time in direction of ω if the driving
errors are assumed to be constant.

The analytic solutions for the position error equation (3.5.13) can again be obtained in a manner analogous to that used for the platform equation. For this, (3.5.13) is written in a new way. Firstly, the term Δg is approximately calculated :

$$\Delta g = g - g^c \approx -\frac{g}{R}\left(\frac{R_o}{R}\right)^2 R + \frac{g}{R^c}\left(\frac{R_o}{R^c}\right)^2 R^c$$

$$\approx \frac{g}{R}\left(\frac{R_o}{R}\right)^2 \Delta R - 3\frac{g}{R}\left(\frac{R_o}{R}\right)^2\left(\Delta R^T \frac{R}{R}\right)\frac{R}{R}$$

(3.6.4)
$$= \frac{g}{R}\left(\frac{R_o}{R}\right)^2 \Delta R - 3\frac{g}{R}\left(\frac{R_o}{R}\right)^2\left(I + \tilde{e}_R^2\right)\Delta R \ ,$$

where

(3.6.5)
$$\left\{ \begin{array}{c} \tilde{e}_R = \frac{1}{R}\begin{bmatrix} 0 & -x_3 & x_2 \\ x_3 & 0 & -x_1 \\ -x_2 & x_1 & 0 \end{bmatrix} , \\[4mm] R = [x_1 \ \ x_2 \ \ x_3]^T , \quad R = \sqrt{x_1^2 + x_2^2 + x_3^2} \\[4mm] g = \text{gravity constant, } R_o = \text{earth radius.} \end{array} \right.$$

Now, (3.5.13) is written

(3.6.6) $\Delta\ddot{R} + 2\tilde{\omega}\Delta\dot{R} + \tilde{\omega}^2 + \frac{g}{R}\left(\frac{R_o}{R}\right)^2 I - 3\frac{g}{R}\left(\frac{R_o}{R}\right)^2\left(I + \tilde{e}_R^2\right)\Delta R = d \ .$

This is a linear differential equation determining the position error. It may be treated by methods of linear system theory but then the solution is difficult to survey. It is more convenient to restrict the problem of error propagation to the important case of latitude-longitude coordinates $\{x_i^s\}$ (Fig. 1) and the constant angular velocity ω^s (3.2.2). Also the distance R of the vehicle from the centre of earth will be approximated by R_0. With the abbreviation

$$\omega_0 = \sqrt{\frac{g}{R}} \qquad\qquad (3.6.7)$$

the specialized position error equation is, in component form,

$$
\begin{aligned}
\Delta\ddot{x}_1 + 2(-\omega_3^s\,\Delta\dot{x}_2 + \omega_2^s\Delta\dot{x}_3) + \left[\omega_0^2 - (\omega_2^s)^2 - (\omega_3^s)^2\right]\Delta x_1 + \\
+ \omega_1^s\omega_2^s\Delta x_2 + \omega_1^s\omega_3^s\Delta x_3 = d_1 \quad, \\[2mm]
\Delta\ddot{x}_2 + 2(\omega_3^s\,\Delta\dot{x}_1 - \omega_1^s\,\Delta\dot{x}_3) + \left[\omega_0^2 - (\omega_3^s)^2 - (\omega_1^s)^2\right]\Delta x_2 + \\
+ \omega_1^s\omega_2^s\Delta x_1 + \omega_2^s\omega_3^s\Delta x_3 = d_2 \quad, \\[2mm]
\Delta\ddot{x}_3 + 2(-\omega_2^s\Delta\dot{x}_1 + \omega_1^s\Delta\dot{x}_2) + \omega_1^s\omega_3^s\,\Delta x_1 + \omega_2^s\omega_3^s\,\Delta x_2 - \\
- \left[2\omega_0^2 + (\omega_1^s)^2 + (\omega_2^s)^2\right]\Delta x_3 = d_3 \quad.
\end{aligned}
\qquad (3.6.8)
$$

Because ω_i^s ($i = 1, 2, 3$) are small in this application compared to ω_0 the position error equations (3.6.8) can be reduced to

$$\Delta\ddot{x}_1 + \omega_0^2\,\Delta x_1 = d_1 \quad, \qquad\qquad (3.6.9a)$$

(3.6.9b) $$\Delta \ddot{x}_2 + \omega_o^2 \Delta x_2 = d_2 \;,$$

(3.6.9c) $$\Delta \ddot{x}_3 - 2\omega_o^2 \Delta x_3 = d_3 \;,$$

which means that the cross-coupling of the position errors can
be approximately ignored. The third equation of (3.6.9) gives
rise to the well-known statement that pure inertial systems are
unstable in earth's radial direction. Therefore, a possible solution for an aircraft is to use altimeter information. Surface
ships, of course, do not have the problem. In the following it
will be assumed that a separate measure of altitude is used for
determining $x_3 = R \approx R_o$ and the magnitude of g . The error
propagation in the level channels (x_1 and x_2 channels) can be
determined without any further specification of the vertical
channel.

The driving functions d_i (i = 1, 2) of the level
channels are determined by (3.5.3) with the acceleration vector

(3.6.10) $$b^s = \begin{bmatrix} \ddot{x}_{10}\delta(t) - 2\omega_3^s v_2^s + \omega_1^s \omega_3^s R_o \\ \ddot{x}_{20}\delta(t) + 2\omega_3^s v_1^s + \omega_2^s \omega_3^s R_o \\ 2(\omega_1^s v_2^s - \omega_2^s v_1^s) - ((\omega_1^s)^2 + (\omega_2^s)^2)R_o - g \end{bmatrix}$$

(note that ω^s = const., $x_3 = R = R_o$ are assumed). Both $\ddot{x}_{10}\delta(t)$
and $\ddot{x}_{20}\delta(t)$ are considered as impulses at launch. The uncoupled
position error equations of the level channels reads now

$$\Delta \ddot{x}_1 + \omega_o^2 \Delta x_1 = \nabla_1 + (K_1^A + K_1^o)b_1^S + \Gamma_1 b_2^S - \Gamma_2 b_3^S +$$

$$+ b_2^S \alpha_3(t) - b_3^S \alpha_2(t) \, ,$$

$$\Delta \ddot{x}_2 + \omega_o^2 \Delta x_2 = \nabla_2 + (K_2^A + K_2^o)b_2^S - \Gamma_3 b_1^S + \Gamma_4 b_3^S +$$

$$+ b_3^S \alpha_1(t) - b_1^S \alpha_3(t) \, .$$

$$(3.6.11)$$

The analytic solution of (3.6.11) is given by

$$\Delta x_i = \Delta x_{i_o} \cos \omega_o t + \frac{\Delta \dot{x}_{i_o}}{\omega_o} \sin \omega_o t + \frac{1}{\omega_o} \int_0^t \sin \omega_o (t-\tau) d_i(\tau) d\tau, \, (i=1,2) (3.6.12)$$

where Δx_{i_o} and $\Delta \dot{x}_{i_o}$ are the initial conditions (3.3.9) and (3.3.10).
The calculation of (3.6.12) leads to

$$\Delta x_1 = \Delta x_{1_o} \cos \omega_o t + \frac{\Delta \dot{x}_{1_o}}{\omega_o} \sin \omega_o t + \frac{1}{\omega_o} [(K_1^A + K_1^o)\ddot{x}_{1_o} + (\Gamma_1 + \alpha_{3_o})\ddot{x}_{2_o}] \sin \omega_o t +$$

$$+ \frac{1}{\omega_o^2}(1 - \cos \omega_o t) \{ \nabla_1 + (K_1^A + K_1^o)\omega_3^S(-2v_2^S + \omega_1^S R_o) +$$

$$+ \Gamma_1 \omega_3^S (2v_1^S + \omega_2^S R_o) - \Gamma_2 [2(\omega_1^S v_2^S - \omega_2^S v_1^S) -$$

$$- ((\omega_1^S)^2 + (\omega_2^S)^2)R_o - g]\} +$$

$$+ \frac{\omega_3^S}{\omega_o} (2v_1^S + \omega_2^S R_o) \int_0^t \sin \omega_o (t - \tau) \alpha_3(\tau) d\tau +$$

$$+ \frac{1}{\omega_o}[-2(\omega_1^S v_2^S - \omega_2^S v_1^S) + ((\omega_1^S)^2 + (\omega_2^S)^2)R_o + g] \int_0^t \sin \omega_o (t - \tau) \alpha_2(\tau) d\tau, (3.6.13)$$

$$\Delta x_2 = \Delta x_{20} \cos \omega_0 t + \frac{\Delta \dot{x}_{20}}{\omega_0} \sin \omega_0 t + \frac{1}{\omega_0} [(K_2^A + K_2^0) \ddot{x}_{20} - (\Gamma_3 + \alpha_{30}) \ddot{x}_{10}] \sin \omega_0 t +$$

$$+ \frac{1}{\omega_0^2} (1 - \cos \omega_0 t) \{ \nabla_2 + (K_2^A + K_2^0) \omega_3^S (2 v_1^S + \omega_2^S R_0) -$$

$$- \Gamma_3 \omega_3^S (-2 v_2^S + \omega_1^S R_0) + \Gamma_4 [2 (\omega_1^S v_2^S - \omega_2^S v_1^S) -$$

$$- ((\omega_1^S)^2 + (\omega_2^S)^2) R_0 - g] \} +$$

$$+ \frac{1}{\omega_0} [2 (\omega_1^S v_2^S - \omega_2^S v_1^S) - ((\omega_1^S)^2 + (\omega_2^S)^2) R_0 - g] \int_0^t \sin \omega_0 (t - \tau) \alpha_1(\tau) d\tau -$$

$$\textbf{(3.6.14)} \quad - \frac{\omega_3^S}{\omega_0} (-2 v_2^S + \omega_1^S R_0) \int_0^t \sin \omega_0 (t - \tau) \alpha_3(\tau) d\tau \;,$$

where $\alpha_i(t)$ ($i = 1, 2, 3$) are the misalignment solutions (3.6.1) of the platform error equation.

Because of (3.6.1), (3.6.13), (3.6.14) the position error propagations of the level channels are

a) sinusoidal with frequency ω_0 and vanishing mean values in consequence of initial position errors

Δx_{i0} ($i = 1, 2$) and initial velocity errors

$\Delta \dot{x}_{i0}$ ($i = 1, 2$),

b) sinusoidal with frequencies ω_0, ω and displaced mean values as a result of the initial platform errors

α_{i0} ($i = 1, 2, 3$),

c) sinusoidal with frequency ω_0 and displaced mean values in consequence of accelerometer bias ∇_i ($i = 1, 2$), scale factors K_i^A and K_i^o ($i = 1, 2$) and nonorthogonality Γ_i ($i = 1,\ldots, 4$) of the accelerometers,

d) time increasing, superposed by displaced oscillations with frequencies ω_o , ω , as a result of gyro drifts ε_i ($i = 1, 2, 3$), scale factors K_i^G ($i = 1, 2, 3$) and nonorthogonality ψ_i ($i = 1,\ldots,6$) of the gyro input axes.

The position error Δx_1 of the first channel is listed as depending on several different error sources :

α) Initial position error Δx_{10} :

$$\Delta x_1 = \Delta x_{10} \cos \omega_o t \ . \qquad (3.6.15)$$

β) Initial velocity error $\Delta \dot{x}_{10}$:

$$\Delta x_1 = \frac{\Delta \dot{x}_{10}}{\omega_o} \sin \omega_o t \ . \qquad (3.6.16)$$

γ) Accelerometer bias ∇_1 :

$$\Delta x_1 = \frac{\nabla_1}{\omega_o^2} (1 - \cos \omega_o t) . \qquad (3.6.17)$$

δ) Imperfect accelerometer scale factor K_1^A :

$$\Delta x_1 = K_1^A \frac{\omega_3^S}{\omega_o^2} (-2 v_2^S + \omega_1^S R_o) (1 - \cos \omega_o t) . \qquad (3.6.18)$$

ε) Constant azimuth gyro drift ε_3 :

$$\Delta x_1 \approx \varepsilon_3 \left(\frac{\omega_o}{\omega}\right)^2 R \left\{ \omega_2^S \omega_3^S \left[\frac{t}{\omega_o} - \frac{\sin \omega_o t}{\omega_o^2} - \right. \right.$$

$$- \frac{1}{\omega} \left(\frac{\omega}{\omega^2 - \omega_0^2} \sin \omega t - \frac{\omega_0}{\omega^2 - \omega_0^2} \sin \omega_0 t \right) \right] +$$

$$(3.6.19) \qquad + \omega_1^s \left[\frac{1}{\omega_0} (1 - \cos \omega_0 t) + \frac{\omega_0}{\omega^2 - \omega_0^2} (\cos \omega t - \cos \omega_0 t) \right] \right\}.$$

Finally, it is emphasized that the above position error propagations (3.6.13), (3.6.14) have the following qualifications :

1. Solution is derived for latitude–longitude coordinate system. Although the error propagation is representative, there are differences from the errors of other systems.

2. A particular flight trajectory has been chosen : Angular velocity data are assumed to be constant ; no vertical acceleration is considered, and horizontal acceleration occurs only at launch.

3. The solution is only good approximation because earth rate coupling has been omitted.

4. Simple models for the error sources were assumed. For instance, accelerometer bias and gyro drift were taken as constant. For the system, analyzed here, this assumption is probably satis factory. However, in applications concerned with Doppler–damped stellar inertial systems, quick alignment, or in–flight alignment, this assumption is not adequate because the random characteristics of the error sources must be taken into account.

References

[1] G.R. PITMAN, JR. (editor) : Inertial Guidance. John Wiley & Sons, New York – London 1962.

[2] C.F. O'DONNELL (editor) : Inertial Navigation Analysis and Design. McGraw-Hill, New York – San Francisco – Toronto – London 1964.

[3] K. REINEL : Allgemeine Fehleranalyse für eine dreiachsig stabilisierte Trägheitsplattform. Deutsche Versuchsanstalt für Luft – und Raumfahrt (DVL), Forschungsbericht DLR FB 68 – 78, Dezember 1968.

Chapter 4
Self - Alignment of Inertial Platforms

4. 1. Introduction

The performance accuracy of an inertial guidance system can be only as good as the accuracy with which the system is initially aligned. For the usual latitude-longitude coordinate system for terrestrial guidance, an error of the initial alignment of the stable platform produces errors in the system's report of distance travelled which vary sinusoidally round about a displaced mean value, as shown in section 6 of chapter 3. This indicates that an important initial condition for an inertial system is the initial platform orientation.

The purpose of this chapter is to explore briefly some of the methods which may be used to align a platform. A mathematical model of the alignment mechanism will be derived. The alignment problem may be considered as the problem of providing coincidence of the platform axes with the computer axes. These two sets of axes can be brought into coincidence by the rotation of the platform axes.

There are basically two types of alignment methods which may be used:

1. The inertial system is aligned with an externally provided attitude reference system.

2. The inertial system, through the agency of its inertial instruments, senses misalignments and automatically aligns itself
to the desired orientation.

In this paper only the second method of self-alignment is considered. In the case of latitude-longitude coordinate system, especially self-alignment methods include leveling
and gyro compassing. Leveling implies the appropriate alignment
of the inertial platform with respect to the local vertical. The
term "gyrocompassing" is used to indicate the appropriate alignment of the platform about the local vertical (that is, azimuth
alignment). Equally, an alternative method of time-optimal align
ment is presented.

This chapter is based on references $[1 - 8]$.

4. 2. Mathematical model

The determination of the orientation of a three-
axis orthogonal coordinate system requires at least two noncol-
linear vectors. For self-alignment the mass attraction vector is
used for leveling, and an angular rate vector, such as the
earth's rotational vector, is used for azimuth alignment. In the
following discussion it will be assumed that the platform initial
ly has been roughly aligned to within a few degrees of the desired orientation so that small angle approximations are valid. It
is also assumed that the gyro input axes and the accelerometer
sensitive axes are coincident and equal to the platform axes $\{x_i^p\}$

(see chapter 3). The last condition will be held because of the faster time of response of the platform stabilization loop. The discussion is generally limited to the case where the computer position errors are essentially zero. This assumption directly implies that the system being aligned is not required to provide navigational information during the alignment process. The acceleration, velocity, and position vectors of the vehicle as well as the angular velocity vector of the reference coordinate system $\{x_i\}$ are assumed to be known. During the alignment procedure they should be constant. Then the platform error equation (3.4.12) in chapter 3 is used as a basis for the following discussion.

The misalignment angles of the platform with respect to the reference system $\{x_i\}$ may be denoted by α_1, α_2, α_3 as shown in Fig. 1. Because of the coincidence of the computer axes and the reference axes the uncontrolled platform error equation is

(4.2.1) $$\dot{\alpha} + \omega \times \alpha = \varepsilon + \tilde{\psi}\omega + K^\varepsilon\omega \; ,$$

see chapter 3, equation 3.4.12. For self-alignment the platform will be influenced by gyro torquing rates $u = [u_1, u_2, u_3]^T$. If we assume that the gyros are perfect instruments, we can neglect the error terms in (4.2.1) which are substituted by the control torques u:

(4.2.2) $$\dot{\alpha} + \omega \times \alpha = u$$

This is the very important equation which describes the time behaviour of the misalignment angles between the platform and reference axes during alignment. The angular velocity ω represents the (constant) rotation of the reference system.

The task of the alignment process is to realize a suitable control law $u(\alpha)$ which reduces the error angles α_i approximately to zero. The establishment of such a control law requires the knowledge of α ,i.e. α_i ($i = 1,2,3$) have to be measured. That is done by the acceleration measurements. By the above-mentioned assumptions the acceleration vector b^P is expressed by the reference acceleration vector b :

$$b^P = (I + \tilde{\alpha})b \quad , \tag{4.2.3}$$

where the coordinate transformation matrix is given by (3.4.1). Because b is taken for granted, the difference between b^P and b will be a measure of misalignment :

$$m = b^P - b = \tilde{\alpha}b = \tilde{b}\alpha \quad , \tag{4.2.4}$$

where

$$\tilde{b} = \begin{bmatrix} 0 & -b_3 & b_2 \\ b_3 & 0 & -b_1 \\ -b_2 & b_1 & 0 \end{bmatrix} . \tag{4.2.5}$$

The problem of suitable alignment control loops is to choose the control torque u as a function of the measurement vector m in an efficient manner :

(4.2.6)
$$\begin{cases} \dot{\alpha} + \omega \times \alpha = u(\tilde{b}\alpha) , \\[2mm] \text{such that} \quad \alpha(t) \longrightarrow 0 . \end{cases}$$

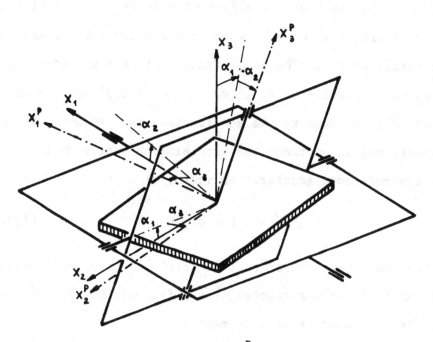

Fig. 1. Coordinate systems $\{x_i\}$ and $\{x_i^p\}$ as well as misalignment angles $\alpha_1, \alpha_2, \alpha_3$.

4. 3. Controllability and observability

The alignment problem (4.2.6) can be solved if and only if the state space equation (4.2.2) and the measurement equation (4.2.4) are completely controllable and completely observable [9] . These two concepts are of fundamental importance to control problems. Controllability is defined as a necessary

and sufficient condition enabling us to design a regulator,
where the task of a regulator is to move an arbitrary initial
state α_o of (4.2.2) to a fixed desired state α_T, which is
usually zero. The observation problem arises in state determination, because the control is a function of the state. In most
practical cases, just as here, we don't know all state variables
of the plant (4.2.2) at each instant. Therefore, it is necessary to determine $\alpha(t)$ from the knowledge of the past outputs
$\{m(\tau): \tau \leq t\}$.

The tests of the algebraic criteria for controllability and observability [9] yields complete controllability
independent on ω and b , and complete observability if and
only if ω and b are not collinear :

$$\omega \neq Kb \qquad (K \text{ arbitrary }) . \qquad (4.3.1)$$

In the special case of a platform, which has to be aligned on
ground before take-off into the latitude-longitude coordinate
system $\{x_i^s\}$ (see Fig. 1 and equations (3.2.1) and (3.2.2) of
chapter 3), condition (4.2.7) stated that alignment is not possible in the close vicinity to the North and South Pole of the
earth.

Because of the satisfactory test for the structural
properties of the alignment system, in the next sections several alignment methods are discussed.

4. 4. Linear control loop (leveling and gyrocompassing)

In this section linear control loops are considered. The control law is put up in a linear manner :

(4.4.1) $$u = Lm ,$$

where L is a suitable chosen linear operator. By (4.2.2), (4.2.4) and (4.4.1) the closed loop alignment system reads

(4.4.2) $$\dot{\alpha} = (-\tilde{\omega} + L\tilde{b})\alpha .$$

Controllability and observability of (4.2.2),(4.2.4) guarantee a suitable operator L such that (4.4.2) is asymptotically stable. In the notation of LAPLACE transformation the solution of (4.4.2) runs

(4.4.3) $$\alpha(s) = \left[sI + \tilde{\omega} - L(s)\tilde{b} \right]^{-1} \alpha_0 .$$

This solution should be discussed in more detail in the special, but important case of ground alignment into the latitude–longitude coordinate system. The fundamental equations are reduced to

(4.4.4) $$\dot{\alpha} = \omega^E \begin{bmatrix} 0 & \sin\varphi & 0 \\ -\sin\varphi & 0 & \cos\varphi \\ 0 & -\cos\varphi & 0 \end{bmatrix} \alpha + u ,$$

$$m = g \begin{bmatrix} 0 & 1 & 0 \\ -1 & 0 & 0 \\ 0 & 0 & 0 \end{bmatrix} \alpha \, , \qquad (4.4.5)$$

$$u = \begin{bmatrix} L_{11} m_1 + L_{12} m_2 \\ L_{21} m_1 + L_{22} m_2 \\ L_{31} m_1 + L_{32} m_2 \end{bmatrix} . \qquad (4.4.6)$$

A structural block diagram of this alignment control loop is presented in Fig. 2. The SCHULER–platform – so called because the stabilization loop of latitude–longitude–coordinate–system–platforms are tuned by conditions according to SCHULER – is actuated by control torques with respect to ω and u and the level accelerometers sense the required measurements

$$\left. \begin{aligned} m_1 &= b_1^S = g\alpha_2 \, , \\ m_2 &= b_2^S = -g\alpha_1 \, . \end{aligned} \right\} \qquad (4.4.7)$$

The LAPLACE equivalent to the differential equation (4.4.2) is

$$s\alpha(s) - \alpha_0 = \begin{bmatrix} -gL_{12}(s) & \omega^E \sin\varphi + gL_{11}(s) & 0 \\ -\omega^E \sin\varphi - gL_{22}(s) & gL_{21}(s) & \omega^E \cos\varphi \\ -gL_{32}(s) & -\omega^E \cos\varphi + gL_{31}(s) & 0 \end{bmatrix} \alpha(s) . \ (4.4.8)$$

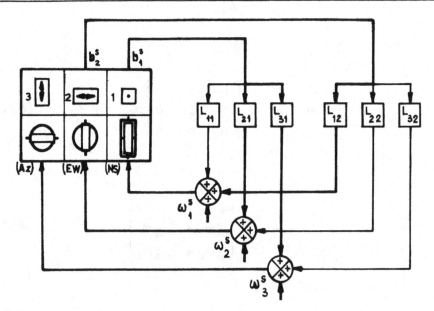

Fig. 2. Block diagram of linear feedback control for the alignment of a
SCHULER-platform.

Usually the elements L_{11}, L_{22}, L_{32} are chosen identically zero.
If in addition the terms gL_{12} and gL_{21} are stated to be much
greater than $\omega^E \sin \varphi$,

$$(4.4.9) \qquad \qquad |gL_{12}| , |gL_{21}| \gg |\omega^E \sin \varphi| \quad ,$$

then the three-dimensional control system will be approximately
decoupled in the so-called leveling and gyrocompass loops :

$$(4.4.10) \qquad \qquad s\alpha_1(s) - \alpha_{10} = -gL_{12}(s)\alpha_1(s) \quad ,$$

$$(4.4.11) \begin{cases} s\alpha_2(s) - \alpha_{20} = gL_{21}(s)\alpha_2(s) + \omega^E \cos \varphi \, \alpha_3(s) , \\ \\ s\alpha_3(s) - \alpha_{30} = (-\omega^E \cos \varphi + gL_{31}(s))\alpha_2(s) . \end{cases}$$

These two uncoupled loops are drawn somewhat thicker in Fig. 2

than the remaining signals.

The level control loop (4.4.10) effects the level-
ing of the x_2^p -axis of the platform (rotation about north–south
axis x_1^p). Often the operator $L_{12}(s)$ is presented by a PI-con-
troller :

$$L_{12}(s) = l_0^{12} + \frac{l_{-1}^{12}}{s}$$ (4.4.12)

which leads to

$$\alpha_1(s) = \frac{\alpha_{10}\, s}{s^2 + g\, l_0^{12} s + g l_{-1}^{12}} \quad .$$ (4.4.13)

This is a damped (oscillatory) motion for $l_0^{12} > 0$, $l_{-1}^{12} > 0$. The
misalignment of α_1 decreases proportionally to $\exp(-\frac{1}{2} g l_0^{12} t)$
$(g_0(l_0^{12})^2 < 4 l_{-1}^{12})$, which leads to a reduction of the error α_{10} to
$\alpha_{10}/20$ within 400 seconds for $l_0^{12} = 1,6 \cdot 10^{-5}$ s/cm [6] .

The gyrocompassing loop (4.4.11) is essential for
azimuth alignment. The azimuth angle α_3 cannot be measured by
the accelerometers and can be only controlled indirectly by the
coupling term $\omega^E \cos\varphi\, \alpha_3$ in the first equation of (4.4.11).
The nonvanishing of this expression $(\varphi \neq \pm 90)$ is equivalent to
the condition of complete observability (4.3.1). The feedback
operators $L_{21}(s)$ and $L_{31}(s)$ are often chosen as P- and PI-con-
trollers :

$$\left. \begin{aligned} L_{21}(s) &= l_0^{21} \quad , \\[2mm] L_{31}(s) &= l_0^{31} + \frac{l_{-1}^{31}}{s} \quad . \end{aligned} \right\}$$ (4.4.14)

Then the solution of (4.4.11) reads

(4.4.15)
$$\begin{cases} \alpha_2(s) = \dfrac{s(s\,\alpha_{20} + \omega^E\cos\varphi\,\alpha_{30})}{s^3 - gl_0^{21}s^2 + \omega^E\cos\varphi(\omega^E\cos\varphi - gl_0^{31})s - gl_{-1}^{31}\omega^E\cos\varphi}\ , \\[4mm] \alpha_3(s) = \dfrac{[(gl_0^{31} - \omega^E\cos\varphi)s + gl_{-1}^{31}]\alpha_{20} + s(s - gl_0^{21})\alpha_{30}}{s^3 - gl_0^{21}s^2 + \omega^E\cos\varphi(\omega^E\cos\varphi - gl_0^{31})s - gl_{-1}^{31}\omega^E\cos\varphi}\ . \end{cases}$$

The motions $\alpha_2(t)$ and $\alpha_3(t)$ will tend asymptotically to zero
if

(4.4.16)
$$\begin{cases} l_0^{21} < 0\ , \\[3mm] 0 < -\omega^E\cos\varphi\,l_{-1}^{31} < -\omega^E\cos\varphi\,gl_0^{21}(-l_0^{31})\ . \end{cases}$$

Also, from (4.4.15) it is obvious that $\cos\varphi \neq 0$ is a
necessary condition for the alignment procedure because of the
integrating factor $\dfrac{1}{s}$ in

(4.4.17) $\alpha_3(s)\Big|_{\cos\varphi\,=\,0} = \dfrac{1}{s}\alpha_{30} + \dfrac{1}{s^2}\dfrac{g(l_0^{31}s + l_{-1}^{31})\alpha_{20}}{s - gl_0^{21}}\ .$

Further types of linear feedback controllers are considered in
detail in [2, 3, 5, 6] .

       Finally, we discuss the influence of constant gyro
drifts ε and constant accelerometer bias ∇ on the steady-state
accuracy of the alignment procedure. From the error analysis of
chapter 3 and the dynamical equations (4.2.2), (4.2.4) follow

(4.4.18) $(sI + \tilde{\omega} - L(s)\tilde{b})\alpha(s) = \alpha_0 + \dfrac{1}{s}(\varepsilon + L(s)\nabla)\ .$

The steady-state error α_{ss} is obtained by the final value theorem,

$$\alpha_{ss} = \lim_{t \to \infty} \alpha(t) = \lim_{s \to 0} s\,\alpha(s) , \qquad (4.4.19)$$

which leads to

$$\alpha_{ss} = \lim_{s \to 0} \{(\tilde{\omega} - L(s)\tilde{b})^{-1} (\varepsilon + L(s)\nabla)\} . \qquad (4.4.20)$$

A short manipulation of (4.4.20) and consideration of the different dimensions of ω^E, l_K^{ij}, ε_i, ∇_i yields

$$\alpha_{1ss} \approx \frac{\nabla_2}{g} , \quad \alpha_{2ss} \approx -\frac{\nabla_1}{g} , \quad \alpha_{3ss} \approx \frac{\varepsilon_2}{\omega^E \cos \varphi} . \qquad (4.4.21)$$

The accuracies of the level angles α_1 and α_2 depend essentially on the bias of the level accelerometers. According to [6] the level error will be about 10 seconds of arc. The steady-state misalignment of the azimuth angle α_3 depends on the latitude φ and is proportional to the drift of the gyro (EW) sensing angular rates about the east-west axis. Assuming $\varepsilon_2 = 0,01°/h$ the misalignment error is about 3 minutes of arc for $\varphi = 45°$. Therefore, the accuracy of leveling and gyrocompassing is influenced mostly by the drift of the east-west-gyro.

4. 5. Time - optimal alignment

In this section a nonlinear control law is considered. If the gyro precessing torques are restricted in magnitude,

(4.5.1) $|u_i(t)| \leq u_o$ $(i = 1, 2, 3)$

we put up

(4.5.2) $u = -u_o \operatorname{sgn} \alpha$,

where $\operatorname{sgn} \alpha$ means a vector whose components are $\operatorname{sgn} \alpha_i$. In
[8] it is proved that (4.5.2) guarantees approximately a time-
optimal alignment procedure. That follows from the facts of
preponderance of the control. Because u_o will be about $10^{-3} \, s^{-1}$
and the system dynamics $-\tilde{\omega}\alpha$ are maximal about $10^{-5} s^{-1}$, the
motion of the alignment feedback loop

(4.5.3) $\dot{\alpha} = -\tilde{\omega}\alpha - u_o \operatorname{sgn} \alpha$

is approximated by the motion of

(4.5.4) $\dot{\alpha} = -u_o \operatorname{sgn} \alpha$.

Therefore, the shortest alignment time will be about

(4.5.5) $T \approx \underset{i=1,2,3}{\operatorname{Max}} \left\{ \dfrac{|\alpha_{io}|}{u_o} \right\}$.

         The control law (4.5.2) is very simple to mecha-
nize by bang-bang controllers and is profitable for time-optimal
alignment. But there is one great difficulty. Our definition
of (4.5.2) assumed that we know what the state of the plant is

at each instant, or in other words, that all the internal variables of the plant can be read out as outputs. By the measurements
m (4.2.4) this assumption is not satisfied. Therefore, in addition to the control law (4.5.2),the regulator must contain another component whose task is state determination. From the definition of a dynamic system, we see that we need two different
kinds of information to determine the state of the plant :

1. Knowledge of the structure of the plant ;
2. Knowledge of the actual inputs and outputs of the plant.

Then we can give a data-processing scheme which
converts these two types of data into a good estimate of the
unknown present state of the plant, if and only if the plant will
be completely observable (see section 4.3 and equation (4.3.1)).
Then we can construct a linear system as an asymptotic state es-
timator [9] or observer [10] for the linear system (4.2.2),
(4.2.4) :

$$\dot{\xi} = D\xi + Fu + Km \; , \tag{4.5.6}$$

$$\hat{\alpha} = S_1\xi + S_2 m \; , \tag{4.5.7}$$

$$\lim_{t \to \infty} \left[\xi(t) - F\alpha(t) \right] = 0 \; , \tag{4.5.8}$$

$$\lim_{t \to \infty} \left[\hat{\alpha}(t) - \alpha(t) \right] = 0 \; , \tag{4.5.9}$$

with

(4.5.10) $DF + K\tilde{b} = -F\tilde{\omega}$,

(4.5.11) $S_1F + S_2\tilde{b} = I$.

The observer (4.5.6) can be considered as a simulation of the
unmeasured part of the dynamic   system (4.2.4), see (4.5.8).
This impression may be heightened by the relation

(4.5.12) $\xi(t) - F\alpha(t) = e^{Dt}(\xi(0) - F\alpha_0)$

as a consequence of (4.5.10). Hence, the state estimation is
asymptotically stable if the eigenvalues of D have only negative real parts.

          The input variables for the control law (4.5.2)
are not the partially unknown state variables α_i ($i = 1, 2, 3$)
but the estimated values $\hat{\alpha}_i$ ($i = 1, 2, 3$). Therefore, the time-
optimal regulator of the alignment procedure is mechanized by
the control law (4.5.2) and the observer (4.5.6 – 4.5.12). In
Fig. 3 the complete control loop of time–optimal alignment is
presented.

          Because the possible observer is not unique, we
require that the dimension of the observer is minimal. Here,
the dimension of the minimal observer is $\dim\{\alpha\} - \dim\{m\} = 1$.
Hence, the matrices D , F and K can be stated by

(4.5.13) $D = [d]$, $F = [f_1\ f_2\ f_3]$, $K = [k_1\ k_2]$.

By (4.5.10) we obtain in the case of a SCHULER–platform (longitude–latitude reference system)

$$f_1 = - \frac{g}{d^2 + (\omega^E)^2} \left[k_1 \omega^E \sin\varphi + \frac{k_2}{d} (d + \omega^E \cos\varphi)^2 \right] .$$

$$f_2 = \frac{g}{d^2 + (\omega^E)^2} \left[k_1 d - k_2 \omega^E \sin\varphi \right] , \qquad \text{(4.5.14)}$$

$$f_3 = \frac{g}{d^2 + (\omega^E)^2} \frac{\omega^E \cos\varphi}{d} \left[k_1 d - k_2 \omega^E \sin\varphi \right] .$$

For $d < 0$ and $f_3 \neq 0$ the minimal observer is asymptotically
stable. Therefore, the unknown variable α_3 of the state vector
can be estimated by $\hat{\alpha}_3$. When the initial condition $\xi(t)$ is
put as

$$\xi(0) = f_1 \alpha_{10} + f_2 \alpha_{20} . \qquad \text{(4.5.15)}$$

(α_{10} and α_{20} are measured by (4.2.4)), and the matrices S_1 and
S_2 are chosen as

$$S_1 = \begin{bmatrix} 0 \\ 0 \\ \dfrac{1}{f_3} \end{bmatrix} , \qquad S_2 = \frac{1}{g} \begin{bmatrix} 0 & -1 \\ 1 & 0 \\ -\dfrac{f_2}{f_3} & \dfrac{f_1}{f_3} \end{bmatrix} , \qquad \text{(4.5.16)}$$

the estimates $\hat{\alpha}_i$ will be

$$\hat{\alpha}_1 = \alpha_1 , \quad \hat{\alpha}_2 = \alpha_2 , \quad \hat{\alpha}_3 = \alpha_3 - \alpha_{20} e^{\alpha t} . \quad \text{(4.5.17)}$$

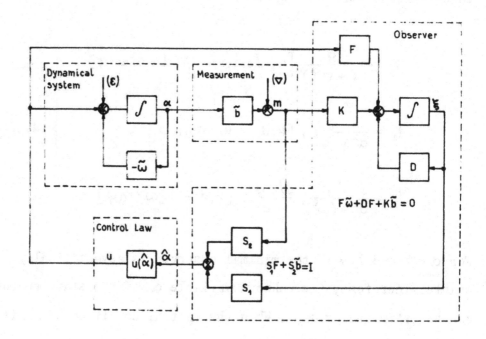

Fig. 3. Complete control loop of time-optimal alignment procedure.

The better the estimation of the unknown state variable, the more stable the observer.

From the regulator, as shown in Fig. 3, the align-ment motions are obtained as presented in Fig. 4 and 5. Depend-ing on the ratio $u_o : d$ the estimation of α_3 will be well-timed or not. Approximately the developments of the misalignment angles are the same as of a directly controlled plant. The pro-

posed nonlinear control loop yields an essentially reduced
alignment time which is about the twentieth part of the usual
times obtained by leveling and gyrocompassing methods.

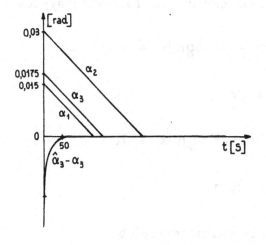

Fig.4. The developments of the
misalignment angles α_i (i= 1,2,
3) and of the difference $\hat{\alpha}_3 - \hat{\alpha}_3$
during time-optimal alignment
procedure ($u_0 = 10^{-4}$ s^{-1},
$d = -0,1$ s^{-1}).

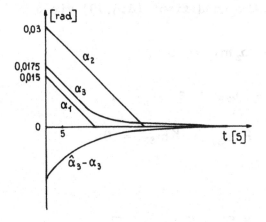

Fig.5. The developments of the
misalignement angles α_i (i=1,2,
3) and of the difference $\hat{\alpha}_3 - \hat{\alpha}_3$
during time-optimal alignment
procedure ($u_0 = 10^{-3}$ s^{-1},
$d = -0,1$ s^{-1}).

Finally, we shall discuss the steady-state error of the non-linear control loop in presence of constant gyro drifts ε and constant accelerometer bias ∇ . From the error analysis of chapter 3 and the essential equations of this section, we have to consider the steady-state response of the following system :

(4.5.18)
$$\begin{cases} \dot{\alpha} & = & -\tilde{\omega}\alpha - u_0 \, \text{sgn}\, \hat{\alpha} + \varepsilon \,, \\[2mm] m & = & \tilde{b}\alpha + \nabla \,, \\[2mm] \dot{\xi} & = & D\xi - u_0 \, F \, \text{sgn}\, \hat{\alpha} + Km \,, \\[2mm] \hat{\alpha} & = & S_1\xi + S_2 m \,. \end{cases}$$

The error α_{ss} after alignment is characterized by

(4.5.19)
$$\dot{\alpha} = 0 \,, \quad \dot{\xi} = 0 \,.$$

Stating $u_{ss} = -u_0 \text{sgn}\, \hat{\alpha}_{ss}$ the conditions (4.5.19) yield

(4.5.20)
$$\begin{cases} S_1\xi_{ss} + S_2 m_{ss} = 0 \,, \\[2mm] -\tilde{\omega}\alpha_{ss} + u_{ss} + \varepsilon = 0 \,, \\[2mm] D\xi_{ss} + F u_{ss} + Km_{ss} = 0 \,. \end{cases}$$

Hence α_{ss} is given by

(4.5.21)
$$\alpha_{ss} = S_1 D^{-1}(K\nabla - F\varepsilon) - S_2 \nabla \,.$$

In the special case of the ground alignment of a SCHULER-plat-form by means of the minimal observer (4.5.13 - 4.5.17) the er-rors will be

$$\alpha_{1ss} = \frac{\nabla_2}{g} \quad , \quad \alpha_{2ss} = -\frac{\nabla_1}{g} , \atop \alpha_{3ss} \approx \frac{1}{\cos\varphi}\left[2\frac{d}{\omega_E^2}\frac{\nabla_1}{g} - \sin\varphi\frac{\nabla_2}{g} - \frac{\varepsilon_2}{\omega_E^2}\right] - \frac{\varepsilon_3}{d} . \right\} \quad (4.5.22)$$

Compared with the steady-state accuracy of the linear alignment loops of section 4 (see equation (4.4.21) the alignment errors of the level angles α_1 and α_2 are the same as before. Equally, the azimuth error is characterized by the drift of the east-west-gyro, but some new effects in consequence of the observer dynamics arise. But if the observer is not too fast, the influ-ence of the errors related to the observer can be neglected. A suitable chosen observer does not cause any loss of accuracy.

References

[1] K. MAGNUS : Kreisel, Theorie und Anwendungen. Springer-
    Verlag, Berlin – Heidelberg – New York 1971.

[2] G.R. PITMAN, JR. (editor) : Inertial Guidance. John Wiley
    & Sons, New York – London 1962.

[3] C.F. O'DONNELL (editor) : Inertial Navigation Analysis
    and Design. McGraw-Hill, New York – San Francisco –
    Toronto – London 1964.

[4] R.H. CANNON, JR. : Alignment of Inertial Guidance Systems
    by Gyrocompassing – Linear Theory. J. Aerospace Sci.
    28 (1961) pp. 885 – 895, 912.

[5] C.S. BRIDGE : Alinement and Long Time Error Propagation
    of Inertial Systems. J. – buch Wiss. Ges. Luft–u.
    Raumfahrt 1963, pp. 211 – 218.

[6] U. KROGMANN : Die selbsttätige Ausrichtung einer Trägheits
    plattform vor dem Start. Luftfahrttechnik – Raum-
    fahrttechnik 11 (1965) pp. 185 – 189.

[7] J.T. KOUBA, L.W. MASON, JR. : Gyrocompass Alignement of
    an Inertial Platform to Arbitrary Attitudes. ARS J.
    32 (1962) pp. 1029 – 1033.

[8] P.C. MÜLLER : Schnelligkeitsoptimales Ausrichten von Träg-
    heitsplattformen. **Ing.–Arch.**40(1971) pp. 248–265.

[9] R.E. KALMAN, P.L. FALB, M.A. ARBIB : Topics in Mathemati-
    cal System Theory. McGraw-Hill, New York 1969.

[10] D.G. LUENBERGER : Observers for Multivariable Systems.
    IEEE Transact. on Automatic Control Vol–AC 11 (1966)
    pp. 190 – 197.

References

[1] K. MAGNUS : Kreisel, Theorie und Anwendungen. Springer-Verlag, Berlin - Heidelberg - New York 1971.

[2] G.R. PITMAN, JR. (editor) : Inertial Guidance. John Wiley & Sons, New York - London 1962.

[3] C.F. O'DONNELL (editor) : Inertial Navigation Analysis and Design. McGraw-Hill, New York - San Francisco - Toronto - London 1964.

[4] R.H. CANNON, JR. : Alignment of Inertial Guidance Systems by Gyrocompassing - Linear Theory. J. Aerospace Sci. 28 (1961) pp. 885 - 895, 912.

[5] C.S. BRIDGE : Alinement and Long Time Error Propagation of Inertial Systems. J. - buch Wiss. Ges. Luft-u. Raumfahrt 1963, pp. 211 - 218.

[6] U. KROGMANN : Die selbsttätige Ausrichtung einer Trägheits plattform vor dem Start. Luftfahrttechnik - Raumfahrttechnik 11 (1965) pp. 185 - 189.

[7] J.T. KOUBA, L.W. MASON, JR. : Gyrocompass Alignement of an Inertial Platform to Arbitrary Attitudes. ARS J. 32 (1962) pp. 1029 - 1033.

[8] P.C. MÜLLER : Schnelligkeitsoptimales Ausrichten von Trägheitsplattformen. Ing.-Arch.40(1971) pp. 248-265.

[9] R.E. KALMAN, P.L. FALB, M.A. ARBIB : Topics in Mathematical System Theory. McGraw-Hill, New York 1969.

[10] D.G. LUENBERGER : Observers for Multivariable Systems. IEEE Transact. on Automatic Control Vol-AC 11 (1966) pp. 190 - 197.

In the special case of the ground alignment of a SCHULER–platform by means of the minimal observer (4.5.13 – 4.5.17) the errors will be

$$\left.\begin{array}{ll} \alpha_{1ss} = \dfrac{\nabla_2}{g} \quad , & \alpha_{2ss} = -\dfrac{\nabla_1}{g} \, , \\[2mm] \alpha_{3ss} \approx \dfrac{1}{\cos\varphi} \left[2\dfrac{d}{\omega_E} \dfrac{\nabla_1}{g} - \sin\varphi \, \dfrac{\nabla_2}{g} - \dfrac{\varepsilon_2}{\omega_E^2} \right] - \dfrac{\varepsilon_3}{d} \, . \end{array}\right\} \quad (4.5.22)$$

Compared with the steady–state accuracy of the linear alignment
loops of section 4 (see equation (4.4.21) the alignment errors
of the level angles α_1 and α_2 are the same as before. Equally,
the azimuth error is characterized by the drift of the east–
west–gyro, but some new effects in consequence of the observer
dynamics arise. But if the observer is not too fast, the influence of the errors related to the observer can be neglected. A
suitable chosen observer does not cause any loss of accuracy.

Contents.